The Origin
of
Matrimony

A Sojourn In
Anthropology

LAWRENCE L. HORSTMAN

Contents

Preface

This book was originally drafted long ago, back when I was a serious student of anthropology. That interest was inspired in my teen years by my brother-in-law, a witty and handsome guy, who instantly became my boyhood role-model, perhaps for the infectious enthusiasm with which he described his studies at the University of Chicago, where he was a doctoral candidate in anthropology. He regaled us with his adventures in the hinterlands of Mexico, tape-recording folk-tales in the native language, and spoke endlessly of the bizarre customs found around the world – and of the bitter quarrels among anthropologists about their pet theories for explaining these things. I hung on every word. In retrospect, a big lesson for me was that it was okay to be scholarly and bookish, even more fun than anything else.

How well I recall my excitement attending a department cocktail party with him and my sister, where the room was abuzz with all the controversial topics of the hour. Some of those people are now big names in the field, but for reasons irrelevant here, Rick's ambitions didn't work out. He died around 1985, alcoholic, broke, and alone.

Naturally, I began reading everything I could lay my hands on about the subject, and when I got to college, took all the courses I could in anthropology and linguistics. (I was a chemistry major, at my father's insistence.) I glowed with pride at getting an A+ from Professor Margaret Mead on a thesis, and then being invited to her office to discuss it, and a possible future in anthropology.

The bottom line is this book, one of a pair, the other on the origin of language, first drafted around 1976. I tried getting it published back then but there were no takers, so it laid in a box, gathering dust. But it was never quite forgotten, for every now and then, I chanced upon this or that article which dovetailed perfectly with my theory. Equally motivating were several alternative theories which got a lot of press, but they were so weak and speculative as to seem ridiculous. How was it possible, I wondered, for such drivel to be taken seriously, or even to get into print at all? It gnawed at me.

When at last I could stand it no more, I dug out those yellowing pages, revised, and updated them and – Ta-Da! – here they are, reborn. If the reader gets half the fun out of reading it that I did in the writing, the effort will have been worthwhile. Enjoy!

1.

The Antiquity of Weddings

1. DISCOVERING SAVAGES. The so-called Age of Exploration is usually dated from the first voyage of Christopher Columbus. It is said to have been fueled by lust for gold and spices, but surely there were other motives, like pure adventure, and besides, Columbus wanted to test his calculations about the size of the earth. He was wrong. Thus do we have "Indians" right here in the USA. As it happened, not much gold was found, but a host of more interesting things were indeed discovered, most notably, a great many people.

They were found all over North and South America, in the frozen Arctic, in the jungles of Africa and Asia, in Australia, and on hundreds of islands scattered over the vast Pacific Ocean. Now, this presented a major puzzle, and all sorts of theories were devised to explain their existence. Were they the Lost Tribes of Israel? If so, did they have souls? Apparently not, for none had ever heard of Jesus, whereupon droves of missionaries were hastily dispatched to baptize them, and apprise them of The Truth. Some Asian peoples were initially confused with orangutans and *vice versa*. Gorillas were first described as a race of hairy humans. Many of the savages ran about stark naked, quite shamelessly at that. It was widely doubted if they spoke a true human language, for they seemed to jabber like animals. Other western sages conjectured that the savages were in fact antediluvian, and spoke the original tongue, before the incident at the Tower of Babel, before the Fall of Man, and so were untainted by sin.

Centuries would elapse before a reasonably coherent understanding of these aboriginal peoples emerged, in the first half of the twentieth century. In essence, it turned out that they were very much like you and I, differing only in culture. That is to say, western history had passed them by. Almost none had any metal tools, and many lived simply by hunting and gathering, without benefit of domestic livestock, gardens, pottery, or writing. Many slept in the open, on the ground.

Accordingly, they were called *primitive* people, a term we shall employ without apology despite its political incorrectness, for alternative terms – euphemisms – are too cumbersome, and not more accurate. Needless to say, however, the word *primitive* is now understood only in the sense of historical and technological developments, not biological.

That is to say, some of our most talented scholars and scientists come from families which, a mere few generations ago, lived naked and knew nothing of the outside world. For that matter, all of us lived that way a few *hundred* generations back. We trust, therefore, that the term, primitive, will no longer be offensive (although as recently as the early twentieth century, it did indeed carry racist connotations). Besides, there are practically none of them left to offend. They are all either dead or have been "acculturated" into the modern world.

2. DISCOVERING MATRIMONY. Among the most surprising discoveries about primitive peoples was that they all got married. No exception was ever found. Accordingly, matrimony is a *cultural universal*, meaning a feature of human culture found everywhere, like fire, weapons, cooking, and language.

Of course, the particular form taken by matrimony varied greatly from people to people. Yet in every case, a socially sanctified ceremony of some kind – a *wedding* – was performed to establish the bond between husband and wife. This fact was not immediately obvious, however, for human societies vary so greatly in structure and family organization that in many, the wedding could easily be missed by the casual observer. The universality of matrimony was firmly established only when anthropologists ventured forth to actually live within the primitive societies, learning their ways by total immersion.

It must be stressed that human marriage is quite different from the life-long pair-bonding seen in many animals. First of all, the pair-bonding seen in animals such as birds is obviously independent of any spoken vows, and unencumbered by externally imposed rules, or threats of punishment for breaking them. Yes, the sin or crime of *adultery* is another cultural universal.

3. THE ANTIQUITY OF MATRIMONY. The universality of matrimony naturally raises the question of when this custom first began. That is the question addressed in this book, along with the how-and-the-why of it. Many have given up on this problem, declaring it unsolvable, but we shall offer a novel hypothesis which seems to me, at least, so simple and compelling that I am amazed that no one has written it before. Of course, it is up to the reader to judge the merits of the solution proposed.

Meanwhile, it has been known for some time that all humans on earth are of a single species and subspecies, and descended from a comparatively small group of "original ancestors" which, according to the Out-of-Africa hypothesis, subsequently spread around the world. For reasons explained later, details of this diaspora remain hazy and controversial, but there is fairly wide agreement that it began roughly 100,000 years ago, give or take, and was largely complete by 60,000 years ago. The figure of 60,000 years is fairly secure, in light of good evidence that Australia was first inhabited at approximately that time. Indeed, the aboriginal Australians not only had matrimony, but one of the most complicated marrying systems in the world, as we shall see later.

This means that the custom of matrimony must be *at least* 60,000 years old, simply because the original migrants who populated the four corners of the world must have already had it. Australia is notable because of its comparative isolation, experiencing little contact with outsiders (compared to Africa, say), yet its aborigines all had elaborate customs of matrimony when first encountered by white people.

Now, 60,000 years may sound like a long time, and certainly it is, considering that the "dawn of civilization" goes back only some 8,000 years, the pyramids of Egypt some 5,000, and the beginning of our calendar a mere 2,000 years. But everything is relative, as it so happens that 60,000 years is only a tiny fraction of the total time – 6 *million* years – that our ancestors have been walking around on two feet.

Several of the leading theories argue that the origins of matrimony date clear back to our very earliest ancestors. In fact, some have argued that the very reason why our original ancestor came down from the trees to walk upright was precisely in order to take care of his "wife," so to speak. Accordingly, we shall spend the next chapter briefly reviewing our ancestry, and the rise of certain social features that may or may not have contributed to matrimony. That will help us to decide which of the many theories, reviewed in Ch. 3, is most plausible, if any. Besides, it is an interesting topic in itself.

4. THE DISCONTENTS OF MATRIMONY. Recent centuries have witnessed dozens or hundreds of social experiments aimed at abolishing matrimony, including numerous "free-love" communes, with or without religious over-tones. Joni Mitchell valiantly sings "we don't need no piece of paper from the city hall keepin' us tied and true." The trendy thing in Hollywood, at least, is to scoff at this ancient tradition. Young men commonly tremble at the thought of marriage; they don't want to get "roped in." Feminists complain bitterly about our sexist patrilineal system (taking the family name of the husband), asserting their heritage, their independent identity by name hyphenations. Other new wrinkles include same-sex marriages, and deliberately unwed motherhood.

This appears to be a growing trend. When I was a kid, vows were sacred and divorce was something you whispered about, like unwed mothers, but now it's all the rage, as easy as trading in your car.

All of these notions and movements will need some rethinking in light of better understanding of how and why marriage originated, and has proved so doggedly persistent. We shall see that our present discontent with this institution is nothing new. In fact, we will suggest that discontent with this institution is exactly the reason why so many variant systems exist! For after all, thousands of societies worldwide were at perfect liberty to abandon this custom altogether, such as when they touched upon some new-found island. They were perfectly free to set things up however they wanted. And yet, every single one of them, without exception, saw fit to retain the custom of matrimony in some form or other. That's food for thought as you venture through these pages.

2.

Looking Backward

We noted that the custom of matrimony must be at least 60,000 years old. However, our ancestors have been walking around for about 6 million years, a hundred times longer. Since several leading theories on the origin of matrimony, to be reviewed in the next chapter, propose that it began very early, even with our earliest ancestors, it behooves us to review some milestones in our evolution, and some key controversies about them.

1. PERSPECTIVE ON DEEP TIME. Six million years seems like a long time but everything is relative. By way of perspective, the dinosaurs appeared around 220 million years ago – let us abbreviate, 220 mya – and were wiped out 65 mya, by an errant meteor. Complex invertebrates such as jellyfish have been around much longer, for at least 500 my,[1] and single-celled life for roughly 2,000 my. So 6 my is not all that long in the big scheme.

We, of course, are *mammals*, and the first of these show up some 180 mya, in the form of small tree-dwelling (arboreal) insectivores, rather like shrews. It was thought that they must have remained small, shy, and nocturnal as long as the dinosaurs were around, but a spectacular fossil mammal recently unearthed and dated 132 mya, had just eaten a dinosaur for lunch.[2] A beaver-like mammal from the middle Jurassic was also recently found,[3] as was another but built for digging, quite like living armadillos or aardvarks.[4] On top of that, a fossil of a gliding mammal was dated to 125 mya,[5,6] possibly an ancestor of bats.

We are also *placental* mammals (eutherians), as distinct from marsupials like opossums and kangaroos, or from monotremes like the egg-laying duck-billed platypus.[7,8] The earliest placental yet found is dated to 125 mya,[9] much earlier than predicted by DNA "molecular clock" methods.

We are also *primates*, the origin and phylogeny of which is a hot topic.[10,11] A mouse-size primate fossil found in China in 2000 was dated to 45 mya[12] and another found in Egypt in 2002 was dated to 56 mya.[13] The closest non-primate relative is said to be the colugos, or "flying lemurs,"[14] which is intriguing in view of the above-cited fossil of a gliding mammal. Living descendants of the earliest primates, called prosimians, are the colorful lemurs and tarsiers of Madagascar, with their gorgeous big eyes and characteristic primate hands. Nearly all now verge on extinction.

We are also higher primates, or *anthropoids*. A variety of apes flourished in Miocene times (5 to 24 mya) but good fossils are rare, most being fragmentary or consisting of just a few teeth. One of the most complete was found in Spain in 2002, dated to 13 mya, and is estimated to have weighed around 30 kg.[15] A number of teeth dated to 10.5 my were reported in 2007, and are said to be indistinguishable from those of modern gorillas.[16] The aptly named *Gigantopithecus* has been known since 1935.[17,18] It stood 9 feet tall, weighed half a ton, and was contemporary with our early ancestors, vanishing only 0.3 my ago, i.e., 300,000 years.

Evidence from DNA and anatomy agree that the chimpanzee is our nearest living relative, and that we shared a common ancestor 5-7 mya – although now they're saying 7-10 mya is more likely.[16] The first fossil chimpanzee was reported in 2005 but was only 0.5 my old,[19] a time when our own ancestors were already well along the yellow brick road.

2. THE FIRST APE MEN. Our ancestors are called *hominids*, although some (but not all) recent authorities prefer *hominins*. How can you tell the difference between a hominid fossil and an ape fossil? You can't, and neither could I. Experts, however, have an uncanny knack for spotting a single tooth or toe-bone, say, and instantly identifying it as hominid or not. Indeed, we are told that fossil teeth sold in Chinese apothecary shops, to be ground up as quack medicine, were a rich source of major finds. The fossil ape, *Gigantopithecus,* was first discovered in that way, by a single tooth, after which more were found at other such shops.[17]

So, how do the experts do it? They do it by the key defining feature of our hominid lineage, which is *walking upright*, called *bipedalism*. No ape does this, or at least, not as we do. It so happens that this change was not a simple matter of standing up tall and walking, but entailed profound skeletal changes, from the toes and ankles through the knees, hips, pelvis and spine, and to the shoulders, neck, and skull. The perfection of these adjustments is nothing short of mind-boggling. For example, it was recently shown that the vertebrae of the human female spine has subtle adjustments of angle to allow her to easily walk upright while pregnant – and that this fine-tuning was already present in some very early fossil hominds.[20] The opening in the base of the skull for the spinal cord, called the foramen magnum, is another marker of upright posture, enabling us to comfortably look forward when standing erect. On that basis, the important Tuang Child was correctly identified as hominid by Raymond Dart, although he endured decades of ridicule for claiming it as hominid. Even the inner ear bones of hominids are distinctive, being modified to accommodate balance when upright.[21]

A second set of distinguishing features are the teeth of us hominids, most obviously, the absence of the large dagger-like canines seen in apes, used by them in their intimidating threat displays. But numerous other and more subtle dental features are also distinctive of hominids.

Most of the earliest known hominid fossils were found only recently, but the majority are quite fragmentary and of uncertain significance. The oldest yet, called *Sahelanthropus tachadensis*, dates to an astonishing 6-7 mya, and raises new questions.[22,23] Others have names like *A. anamensis,*[24] *Orrorin tugenensis*, and two species of *Ardipithecus.*[25,26] The latter, *Ardipithecus ramidus* – "Ardi" – was discovered in 1994 and dates to 4.4 mya, but took 15 years to piece together; Ardi was named "Breakthrough of the Year" for 2009.[27] However, the Lucy type offers the best insights for our purposes.

3. LUCY AND THE NUTCRACKERS. At the time this book was first drafted, a great discovery had just been made, and her name was Lucy. She was then the oldest fossil of a fully bipedal hominid ever found, and lived from about

4 mya to about 3 mya. Her proper name is *Australopithecus afarensis*, after the Afar region of Ethiopia where she was found. The easier name, *Lucy,* was given by her discoverers during an all-night celebration with a tape-player blaring *Lucy in the Sky with Diamonds*, under the African stars. Her finding was a combination of luck and preparedness by the then-young Donald Johanson, memorably told in his book.[28] His good fortune continued to the very next season, when his team found more of Lucy's kind, dubbed "Lucy's family." They probably drowned in a flash-flood. Several more have since been found, some taller, making *A. afarensis* the best known of the very early hominids.

What was most revolutionary about Lucy is that her body and brain were no bigger than a chimpanzee, yet she was almost fully adapted for upright walking. Thus, brain size – presumed to equate with "intelligence," whatever that means – increased only later, after adopting bipedalism, and therefore could not have been the "cause" of upright walking. (It was previously believed that bigger brains must have come first, to enable sufficient intelligence to cope with fearsome predators.) A few years later, her upright walking was established beyond doubt, when Mary Leakey's team discovered fossil footprints in what was once freshly-fallen volcanic ash, fitting Lucy's foot and dating to the same time.[29] Lucy had an almost fully modern foot and stride, even though her arms, hands, and shoulders still had ape-like features.

Another revolutionary upshot of Lucy's discovery was the realization that she came before, rather than after, another group of hominid fossils, *Australopithecus robustus,* so called because of their very husky, almost gorilla-like physique. It is now believed that *A. robustus* was a side-branch which must have split off from the Lucy type, and then became increasingly gorilla-like. Indeed, late specimens of them, such as *Zinjanthropus*, a.k.a. *A. bosoi*, even had the crested skull needed for the attachment of powerful chewing muscles, seen also in gorillas, giving rise to its nick-name, Nutcracker Man. These observations were said to imply a diet similar to gorillas,[30] but someone has suggested they could just as well have dined on crabs and clams.[31] One authority recently proposed that the robust hominids also had a social structure like gorillas, that is, "harems" dominated by a silver-back alpha-male, on the basis of the smaller size of *A. robustus* females (sexual dimorphism). Other hominid species, including ourselves, do not show marked sexual dimorphism. In any case, the Lucy type must have coexisted with the Nutcrackers for a long time, since remains of both are found which date to similar times.

4. WHY WALK? With that bit of background, we come to the first major question related to our thesis: why did our ancestor come down from the trees? The answer to this question has important bearing on everything to come, but is almost as controversial as the origin of matrimony. Many of the current theories border on the ridiculous when compared to the most obvious solution, first proposed by Charles Darwin: *our ancestor had learned to hunt!* That is to say, he – or perhaps she - had discovered that by throwing rocks or by thrusting with a pointed stick, he could drive away predators, and even stun or wound small game. This would explain a great deal about the development of both human society and anatomy.

Support for this hypothesis has grown stronger with recent discoveries in primatology. Jane Goodall demonstrated in the 1960s that chimps use tools, specifically, a stick for extracting termites to eat, and commonly throw things at various adversaries, although not usually with very good aim or much effect.[32] It is also known that chimps periodically hunt and eat meat, such as green monkeys. More recently, it has been shown that chimps in the wild invented the custom of cracking open nuts by placing them on a stone and striking them with another stone, which practice subsequently spread to neighboring groups.[33] In laboratory studies, it was demonstrated that they actually think about and plan their use of tools.[34] More astonishing, and of obvious relevance to the hunting hypothesis, chimps in the wild were recently observed to use spears, sharpened with their teeth, to impale small animals![35] Curiously, this was observed only in females, perhaps being their easier way of getting meat since males do not share their kills with females.

All of this has led to claims by numerous investigators that chimps have a kind of true "culture," similar in essence to human culture.[36-40] This, however, is controversial, since it depends on how you define "culture," a topic we shall touch upon later. Much of the contentiousness of this issue stems from our scientific tradition of regarding animals as utterly different from humans, a tradition which clearly stems from the Christian tradition that humans alone have souls. Jane Goodall remarked at a meeting that back in the 1960s, it was forbidden to even hint that chimpanzees might have minds.[41] They were assumed to be robot-like mechanisms. In any case, enough has been said to make it perfectly reasonable that the original hominid ancestor, having discovered simple weapons, began walking and running in order to free the hands for carrying and using the weapon, as Darwin proposed.

5. THE SCAVENGING HYPOTHESIS. That brings up a once-respected alternative theory, that our early ancestors were mainly scavengers, eking out a living by furtively snatching leftovers from large predators, or already-dead carcasses.[42] The assumption here is that our early ancestors were weak and practically helpless among the big cats, and needed bipedalism mainly to scamper away from them. But at least one documentary film records a group of just three bushmen of the Kalahari Desert – slim and small-boned people, at that – armed only with wooden spears, facing down and driving off a snarling lion intent on stealing their kill. So much for the helpless scenario. Even chimpanzees are quite fearsome foes, but imagine what they could do if armed with spears.

Once initiated, the hunting habit would have rapidly led to increasing adaptedness for throwing and running, by a combination of natural selection, inter-group competition, and sexual selection for mates with good skills at the new art. Consistent with this is another defining feature of our lineage: superb adaptation for long-distance running.[43] Here again, the reason for this has been considered a mystery, but in my not-so-humble opinion, it's a no-brainer: our ancestors had got in the habit of chasing down prey that they had only wounded. Our uniquely abundant sweat glands and absence of body hair contribute to this ability by enabling us to rapidly shed heat. Believe it nor not, over a long haul, sometimes spanning days, a fit human can run down to exhaustion almost any animal, even if only lightly wounded, a method fittingly called "exhaustive pursuit" and well-documented among living peoples.

Related to all of this is the matter of the meat diet. Living apes are largely vegetarian, only occasionally hunting meat. Our kind, on the other hand, eats meat regularly. The great advantage of this is seen simply by comparing the eating habits of apes to carnivores. The former are busy foraging for fruits and leaves all day long, whereas the latter must eat only every few days, spending the rest of the time snoozing. It's a simple matter of proteins and calories.

There's more. As we shall presently see, the hunting hypothesis can also account for certain key features of human social structure, relevant to matrimony. But first, let us briefly review the more fashionable theories.

6. MORE FASHIONABLE THEORIES.

At least two respected theories hold that our first ancestor came down from the trees in order to get married, so to speak. Leaving that for the next chapter, it was recently noticed that orang-utans often move through trees in a kind of walking motion, leading to the hypothesis that upright walking developed in that way.[44] However, strong criticisms of that notion appeared in a subsequent issue, and in several articles where other experts were interviewed,[45] the chief objection being well-established evidence that bipedalism evolved from a knuckle-walking ancestor,[46] and that our closest living relative is the knuckle-walking chimpanzee. Orangutans are genetically more distant. Of particular interest to us is the editorial accompanying that article, which provides a thumbnail list of what appear to be the leading theories:

> "To date, there is no consensus about the adaptive scenario that could have led to the adoption of terrestrial bipedalism. Many theories have been proposed, including the postural feeding hypothesis [K.D. Hunt]; a model [C.O. Lovejoy] attributing bipedality to the social, sexual and reproductive behavior ...; the thermoregulatory hypothesis [P.E. Wheeler] ... to reduce the amount of the body directly exposed to sunlight ...; and the appeasement model [N.G. Jablonski and G. Chaplin], which focuses on bipedal displays that allow for the relatively peaceful resolution of conflicts." [44]

Conspicuously absent from this list is the hunting hypothesis.[47] Also missing is the scavenging hypothesis,[42] and the variant know as the "sentinel hypothesis" – that standing up was needed to see over tall grass to look out for trouble. Far-fetched? Your call. So, why has the hunting hypothesis fallen into such disfavor? The only explanation I can think of is consistency with current paradigms about how evolution works, which still reigns despite findings showing otherwise. In brief, it holds that everything happens by genetic mutations. The hunting hypothesis for upright walking differs in holding that the key event was the *invention of a new strategy* – namely, weapons. No high intelligence was required, only hands, and a flash of insight. However, the advent of weapons created new social dynamics that led to the rise of the "higher intelligence" that we so proudly lay claim to.

We are not entirely alone in favoring the hunting hypothesis. Washburn, for example, embraced it in 1978, when it was already unfashionable.[48] More recently, J.D. Speth, in the closing paragraph of his book review of *The Evolution of Human Hunting,* expresses strong reservations about rejecting the hunting hypothesis:

> "This is perhaps also the time to inject a note of caution. Though there have been a number of provocative and convincing arguments ... that suggest we have overemphasized man's organized, technologically aided hunting prowess in the Pleistocene, there is now a stampede towards the opposite pole, to view premodern humans as essentially opportunistic scavengers who lacked 'planning depth,' ... who wandered 'irregularly,' almost dumbly, over the landscape in search of food. I fear that the pendulum is swinging much farther and faster than either current theory or data justify, and we will find ourselves a decade or so down the line wondering how we could have ever have been so naive or blind." [49]

"Naive or blind" – he said it, not me – sounds about right.

7. WHY EGALITARIAN? The hunting hypothesis can also explain certain pivotal features of human society. By way of background, all known hunter-gatherer societies were *egalitarian,* meaning that they had no chiefs or kings. Instead, they decided matters collectively and democratically. This stands in sharp contrast to other higher primates, nearly all of which are ruled by an alpha male. Bullying and intimidation are almost constant facts of daily life. And so, the question arises, *how and why did our ancestors become egalitarian?*

In light of the above, the answer is obvious. The advent of weapons spelled the doom of dominance by alpha males and pecking-orders. To see why, consider that even a scrawny youth or dainty female can plunge a spear into the heart of the strongest male – perhaps from behind, or while he is sleeping. Indeed, it seems highly probable that such murders actually occurred many times before the message finally came across: *that groups must live amicably together or face murder.* It is not for nothing that weapons are popularly known as "equalizers." It was weapons, not intelligence, that set us 'above the beasts'.

The same phenomenon explains the recession of the dagger-like canine teeth used by apes in their threat displays of intimidation, but characteristically absent in hominids. Such displays are obviously open provocations in the eyes of an armed adversary. A combination of 'natural selection' (frequent death by murder) and sexual selection – distaste for mates with big canines – would have eliminated them. Thus, the advent of weapons, which made murder as easy as slicing a pie, would have had the inevitable result of calming too-quick tempers, and demanded an egalitarian social order as *the only solution*.

8. MEET GENUS *HOMO*. By about 2 mya, the australopithecines had waned away and the first of our genus, *Homo*, began to appear, in the species, *Homo habilis*. These fossils are decidedly less ape-like than Lucy or the others, more human, and with a somewhat larger brain. The first of these to be found, by L.S.B. Leakey, was named *H. habilis*, meaning "handy-man," because of the crafted stone implements found with it. These are the earliest known stone tools, termed *Oldowan* style for the region where they were found.[50,51] Several additional specimens are now known, one of the best being KNM-ER-1470, originally named *H. rudolfensis* but now assigned to *H. habilis*.[26] The brain size of *H. habilis* was 600-775 cc, significantly larger than its predecessors.

H. habilis marks the beginning of crafted stone tools, a watershed development which must have culminated a long prior history of using non-crafted stones, found on the spot and used as-is: rocks, clubs, sticks, or spears, sharpened as female chimpanzees do. (Incredibly, however, some authorities deny any use of weapons or tools prior to those that were obviously crafted of stone.) The crafting of stone tools was probably related to some of the new anatomical features, including the somewhat larger brain, for this entailed the discovery that by banging one rock against another, sharp cutting edges and points could be produced, provided the right kind of stone was used. This is perhaps the first development that might plausibly be linked to a requirement for increased intelligence.

Not long after *H. habilis*, the next later species began showing up, first found in Europe and Asia, now called *H. erectus*. Famous examples include the Heidelberg Man and Peking Man. Erectus endured for about 1 my, up until the rise of our modern species, *H. sapiens*, about 0.3 mya

(300,000 years). The African version is called by some *H. ergaster*. The most complete specimen yet found is the Nariokotome adolescent boy, a.k.a. WT 1500A[52], from Africa, dating to 1.6 mya. A lot happened during the reign of *H. erectus / ergaster*, four things in particular.

First, most of the nearly 3-fold increase in brain size took place during *H. erectus*. Second, stone tools became ever more refined, in a series of steps or stages. Third, mastery of fire was achieved. Big-game hunting is fairly well established within this 1.3 my period (though it may have begun earlier or, as some insist, only later). Fourth, *H. erectus* is the earliest hominid found outside of Africa, i.e., in Europe and Asia. We now consider some of these developments and their implications for the rise of societies and families as we know them.

9. FIRE AND COOKING. The advent of fire-handling was certainly a milestone and, needless to say, was a prerequisite for cooking. Cooking is now a cultural universal, meaning that all societies ever found cooked at least some of their food. Unfortunately, however, assigning a date to the advent of fire-handling has proved contentious. Raymond Dart in the 1940s gave the name, *A. prometheus*, to one of his specimens, after the legendary stealer of fire, because of bones that appeared to be burnt; but it was later found that the burnt appearance of the bones, now dated to 2.8 mya, was due to mineral staining, not fire.

For a long time, the earliest accepted evidence of fire-handling was at the site of Peking Man, an *H. erectus* dated to 0.5 my (500,000 years, or 500 kilo-years, ky) but more recent analysis has cast doubt even on that.[53] In 2004, a claim was made for fire-handling dating back to 790 ky,[54] the time of *H. erectus*, but not all accept the evidence given, for the simple reason that it is exceedingly difficult to clearly distinguish between signs of natural fire and man-controlled fire.

The same is true for evidence of cooking, i.e., charred bones that show signs of butchering or crushing to extract marrow were not necessarily deliberately cooked, they may have been burned naturally at a later time. Thus, purported evidence of man-controlled fire at Baringo basin of Kenya (1.4 my) is viewed with great skepticism.

At a symposium, Richard Wrangham's theory of cooking was hotly debated.[55] Briefly, he argues that cooking had a big survival advantage by reducing the energy needed for digestion, thereby allowing the spurt in growth of the human brain that took place between 1.9 my and 0.3 my, and to fuel the human brain's exceptionally large energy demands. That idea is not new, but Wrangham presented a lot of animal studies and other data documenting the energy efficiency of cooked meat. Many other theories have been proposed to explain – or to rationalize – the advantage of cooking, such as reduced infection by parasites and other pathogens. On the other hand, some have noted downsides, such as the destruction of certain vitamins by cooking.

The problem with Wrangham's theory is absence of solid evidence of fire or cooking that long ago, 1.6-1.9 my. The first *undisputed* evidence consists of fire sites encircled by hearth stones or bones, but these sites date to relatively recent times, 250 - 300 ky, when modern man was emerging and already had the fully modern brain size. The eminent paleontologist, C. Loring Brace, while admitting that fire handling may go back as far as 800 ky, argues that cooking did not start until 250-300 ky, pointing to a reduction in size of the lower face of *Homo* around that time, taken as evidence of reduced need for strong chewing muscles. Recently, still another theory has gotten some press, namely, that the cooking of starchy roots and tubers rather than meat was the energy-rich fuel that propelled the growth in brain size.[56]

What are we to make of all of this, *vis a vis* the matrimonial family? Not much, to be perfectly frank. However, it does seem reasonable to suppose, in view of the female's role as mother, that there was an early division of labor in which the males did the hunting and the females stayed near home-base, whether or not she was busy cooking.

10. FOOD SHARING. Closely related is the matter of food sharing. It is widely accepted that group-sharing of food is a human cultural universal, distinct from the habits of apes. That is to say, human hunters are unique in returning to a base camp carrying their kills to be shared, another reflection of the egalitarian social order. The late Glynn Isaac claimed to see evidence of this behavior at several Leakey sites in northern Kenya and Ethiopia.[57] The evidence for a fixed camp consisted of many worked stone implements in a region of sandy soil, indicating that the stone used to make the tools was carried there. Numerous animal bones were also found bearing marks of cutting and crushing, including those of large animals, even a hippo. Leakey called such sites "living sites" and Isaac argued that this was evidence of food sharing at fixed sites among those early hominids. Others, however, were unconvinced, arguing that the bones were not necessarily hunted and killed, they may have accumulated there by natural processes, and some show tooth marks of predators.

The bottom line, as usual, is that the factual evidence, in the form of concrete artifacts, is equivocal, a matter of interpretation, open to doubt. On the other hand, granting the hunting hypothesis, and the egalitarianism that must have soon followed, then the sharing of food seems like a necessary corollary.

11. BLUSHING BRIDES. When it comes to the fine details of early hominid social structure, the fossil record is silent. On the other hand, quite a lot can be inferred just by looking at your fellow humans. Humans can smile and laugh, for example, and weep in sorrow. Indeed, our range of facial expressions is vastly greater than any other animal, expressing such feelings as shame and embarrassment, grief, joy, sorrow, and a thousand other nuances of vastly greater range and subtlety than any ape could manage. Among other things, apes simply do not have many of the specialized facial muscles that control these expressions. Only humans can cry in sadness, or in joy, and blush in arousal or in shame.

It thus appears that hominids are continuing a clear evolutionary trajectory of increasingly subtle and intimate communicational displays, evident first in mammals compared to reptiles, and now in man compared to apes – a kind of window into one another's heart, opening ever wider as we evolve. These observations inform us of a very long period in our history when human social groups must have been extremely warm, intimate, tender, and highly responsive to one another's expressions, for otherwise such qualities could not have arisen.

Human sexuality is also unique, the female being sexually ready at all times, not just in oestrus (ovulation), and her breasts advertise her sexual appeal even when not lactating. It is said that bonobos, a kind of pygmy chimpanzee, have somewhat similar sexual habits, but details are controversial and seem to depend on the impressions of particular observers.[58]

(We prefer to avoid altogether the question of the alleged human propensity for violence, and whether this was inherited from apes. It took me a lot of reading to finally realize that the primatology literature is largely irrelevant to the human condition, i.e., people are not apes. To say more would lead us into a quagmire of sociobiology, theory of instincts, psycho-behavioral genetics, concepts of how evolution works, and other areas of junk science best ignored here.)

12. SAYING "I DO." It is obvious that you cannot have matrimony without language to explain what's going on, the associated rules, uttering the vows, and so on. Therefore, language is an essential prerequisite, and must have arisen before matrimony. (On the other hand, a number of theories considered in the next chapter argue that humans bonded as couples long before the rise of formal matrimony.) In any case, there is no generally accepted hypothesis for when language arose, or how it arose, or even exactly what it is – although in a sequel to this book it will be shown that it was another overnight invention, analogous to the invention of weapons.

Efforts have been made to link certain features of fossil hominids to the rise of language. One concerns skeletal features for attachment of breathing muscles that we use in speech, said to go back 1.8 my, and a related feature is the size of the spinal canal, to carry the extra nerves needed for fine breath control in speaking.[59] Other skeletal features include points of bone in the skull for attaching muscles that we humans uniquely possess for controlling the tongue. Another is the shape of the skull itself, raised in us to allow space in the roof of the mouth for a pharynx, giving us a high forehead. The appealing theory of Lieberman and Crelin, who argued on this basis that Neanderthals could not talk very well, was based on mistaken ideas.[60,61] However, none of the above – or others such as the FOXP2 gene – offer convincing evidence for the time of the origin of human language.

We shall return to this topic later in the book, and will also later continue the more recent history of our species, the last quarter-million years, because our hypothesis suggests that the origin of matrimony was closely linked to the final surge of our species toward modern times.

3.

The Usual Theories

In this chapter, we briefly review the main theories of matrimony, old and new, by way of assessing the competition. If the reader thinks that any of them are really good, then he need read no further, for the problem is solved. But if he finds them less than compelling, then later chapters will be of interest.

1. THE 19TH CENTURY. It was in the 19th century that something like a 'science' of ethnology (cultural anthropology) began to take shape. Notice, by the way, the two distinct branches of the field, *physical* anthropology, including *paleo*-anthropology, as in the previous chapter, as distinct from *cultural* anthropology. The quality of the work back then was very uneven, mostly based on reports of sea captains, fur-trappers, traders, explorers, colonial administrators, settlers, missionaries, and miscellaneous adventurers. These reports, accurate and not, filtered back to Europe where they were collected and much discussed by 'armchair ethnographers' – who promptly set about devising all sorts of theories about the primeval state of humanity, and the origin of marrying.

There were some remarkable exceptions who produced work of enduring quality, two in particular. Lewis Henry Morgan, a lawyer in upstate New York, was one, now often called the father of American anthropology. He was fascinated by the local Iroquois "Indians" (better called *Amerindians*), and it is easy to see why. When he inquired about their words for relatives such as our uncles, aunts, cousins, in-laws, and so forth, he noticed that their terms were often completely different from ours, and made distinctions that we do not, such as whether the uncle or aunt is on the mother's or father's side, and whether the cousin is the child of the mother's sister or mother's brother (or father's sister or brother), and so on.

He wrote up his findings on the Iroquois, published 1851,[1] then expanded his study, comparing Iroquois kinship terms and marrying practices with others in the region, and then, all over America. Finally, he mailed inquiries to colonialists all over the world, ultimately resulting in one of the finest early inventories of such customs.[2] Now, it just so happens that these kinship terms are closely related to marrying practices and social organization. For example, in many clan-type societies, everyone in your clan of the same generation is a 'sister' or 'brother', and you must marry into a different clan, otherwise you would be committing incest, even though many of these 'sisters' and 'brothers' are only remotely related to you. Conversely, certain cousins are often *preferred* as spouses. And there were other surprises, too, such as the fact that many primitive societies were matrilineal, meaning that the married couple took the 'family name' of the mother, quite the opposite of our own patrilineal tradition.

Leaving the details for later, the bottom line is that all of this information proved irresistible to the armchair ethnographers, tempting them to devise a variety of theories about the hypothetical *original* society – and how and why marrying began. The basic premise was that the world's societies reflect different stages or phases in the evolution of cultures, and that they could be ranked and classified accordingly. This was not entirely unreasonable, for after all, Charles Darwin had demonstrated in 1859 that some animals are more primitive than others (arose earlier), yet lived on. Why shouldn't the same be true of human societies?

Thus do we have, for example, the theory of Bachoffen (1861),[3] proposing that women in the primeval condition must have been constantly victimized by the lust and tyranny of males, and therefore devised a clever strategy for subduing the masculine gender. How? By inventing religion, of course! Accordingly, he sought to demonstrate that the ruling deities were originally feminine: moon over sun, earth over sky. Exalted womanhood ruled things here on earth, too. Most enduringly, it was she who establishes the family, and demanded spousal fidelity by the matrimonial vows. The inspiration behind this 'primeval gynocracy' was the discovery that some societies were indeed matriarchal and matrilineal and, perhaps because such societies are rather rare, and contrast so sharply with our own, he thought they must reflect an early stage of humanity. By the way, he and others of his era were convinced that the myth of the Amazonian gynocracy was literally true.

Interestingly, a somewhat similar hypothesis was more recently advanced by the eminent poet and scholar, Robert Graves, in *The White Goddess,*[4] based on a combination of intriguing archeological finds (small carved 'sex goddesses' widespread in the Pleistocene), and ancient mythologies in which the moon goddess – the *White Goddess* of his book title – evidently stool at the pinnacle of the pantheon. His book spawned a whole genre of lesser works by assorted feminists wistfully romanticizing the bygone era of female rule, beginning perhaps with E.G. Davis' book,[5] which opens on a promisingly sober and scholarly note, but soon slides into fantasies of glorious blond Celtic warrior women galloping across the moors on white stallions, swords on high.

The next step in Bachofen's scheme was the revolt of the males, overthrowing the benevolent gynocracies of yore, accomplished by the male's invention of a higher and more spiritual religion. Mixed in with his account is an explanation for numerous customs described in living societies. For example, in addition to matrilineal descent, he made much of the *couvade,* being the custom in some societies of the father taking the newborn child to bed and pretending to give birth to it, accompanied by a series of rituals. According to Bachofen, that was another survival of the stage in which males were attempting to assert their importance by pretending to be mothers.

By a curious coincidence, in the very same year that Bachofen's *Mother Right* was published, Sir Henry Maine's *Ancient Law*[6] appeared, arguing the exact opposite: that the original society must have been patriarchal and patrilineal. At about the same time, the Edinburgh lawyer, John McLennan, was reconstructing things in his own way, also drawing the conclusion that the primeval arrangement was patriarchal and patrilineal. To glimpse the logic that set him off, he was intrigued by reports of the killing of female babies (female infanticide), which McLennan took to be a survival of an ancient custom, explained by reasoning that females were of lesser importance in the struggle for survival, so the excess were killed, making fewer mouths to fill.

(Later studies revealed that infanticide was not a normal or widespread practice, and was not limited to females, having been instead a response to the horrors visited upon them by the white people.[7] On the other hand, even chimpanzees are observed to practice infanticide, believed to be a measure of population control.[8] They are also occasionally cannibalistic. In parts of aboriginal Australia, infanticide sometimes reached 30%, probably in response to often scarce food resources.)

Infanticide was woven into McLennan's account of the 'primal horde,' arguing that as a result of this practice, women were in short supply, thus requiring that several males share each female. This was tendered to explain reports of polyandry (multiple husbands). By the same stroke, the scarcity of women led to frequent raids on neighboring groups to capture females, whereupon it presently became a matter of honor that one's woman be stolen from another. That is to say, it became unmanly to take a woman from one's own group, then sinful. This was supposed to account for *exogamy*, being the requirement to marry outside of one's own clan. The term, *exogamy*, was coined by McLennan.

McLennan went on to explain the transition from the original polyandry to the more common polygyny (multiple wives). His rationale was that a particularly successful raider would accumulate too many wives to feed, so he shared his harem with his brothers. Later on, we are told, the brothers could share only after the owner's death. This was purported to explain the many known cases of a brother's obligation to marry or care for the widow, a custom known as the *levirate*. His theory goes on to explain how these circumstances led to the reckoning of kinship ties, exogamous clans, the rise of patrilineal marriage, totemism, and many other curious features reported among various societies.

Indeed, it looks like McLennan was trying to explain just about every type of primitive society and practice then known, as if each had survived to reflect a distinct stage in cultural evolution. The big problem with his theory was that it was not demonstrably any better than a half-dozen others, each of which was entirely different, or even exactly opposite. This came to light in his famous dispute with Morgan, who by that time (1877) had himself succumbed to the temptation to write a theory.[9]

Morgan classified primitive societies into several stages, first *savagery*, then *barbarism*, and finally *civilization* – like his own, in upstate New York. Each was divided into a lower, middle, and upper phase. At the origin of it all was a hypothetical *consanguinous* phase in which brothers married their sisters. This was followed by a *punaluan* period, a form of group marriage in which siblings were forbidden to marry, and then, after a transitional stage, a patriarchal stage. Last to arrive, of course, was that epitome of real civilization, monogamous marriage. Fanciful as this succession may now seem, and indeed was, it was all worked out in meticulous detail, and accorded well with Morgan's extensive factual knowledge of primitive societies.

The fatal flaws in all these theories surfaced when their author's began squabbling with one another about who was right. Notable among these quarrels was that between Morgan and McLennan. At first, McLennan admired Morgan's work and they corresponded amicably, but this was not to last. Morgan's *Ancient Society*[9] devotes considerable space to attacking McLennan's theory. Unfortunately, however, Morgan's theory – or at least, those parts of it dealing with the successive stages of culture and matrimony – was scarcely any better, and was later shown to be riddled with factual errors, too. For example, he assigned Hawaiians to his "middle savagery" epoch because they lacked the bow and arrow (and pottery), but we now know that Hawaiians arrived on the islands quite recently, bringing with them an aristocratic type of society (kingship) with roots in Asia. By the way, the king and queen there were brother and sister, a practice seen also among the Inca royalty.

The bottom line is that Morgan's theory cannot be taken any more seriously than any of the others of the 19th century. It was really just a lot of yarn-spinning, and all of them were thrown overboard in disgrace when Franz Boas seized the helm of anthropology. Unfortunately, the factual contribution of Morgan – his great catalog of kinship terms – was forgotten, until rediscovered decades later by Leslie White.

2. FREUD'S THEORY. Sigmund Freud's theory was immensely influential, not only because he is the father of psychoanalysis, but also or mainly because he was such a captivating writer that his *Totem and Taboo*[10] of 1919 jumped to the top of the best-seller list. Indeed, he was nominated twice for a

Nobel Prize, but for literature, not science. This must have rankled him, for he imagined himself a paragon of the Scientific Method, and was openly contemptuous of religion, as seen most clearly in his *Future of an Illusion*.

Aside from being a scholar of classics and mythologies, he was also an avid armchair-ethnologist, familiar with all of the above early theories. Not surprisingly, therefore, his theory was very much in that tradition except, of course, that he tailored it to his notion of the *Oedipus complex*. As everybody nowadays knows, this complex is named for the Greek play, *Oedipus Rex*, in which King Oedipus romances Jocasta, only to eventually discover, to his horror, that she was his own mother. Mortified, he plucked out his own eyes in a fit of shame and disgust. The gist of the complex is that boys harbor secret sexual desire for their mothers but are frustrated by fear of the father's wrath. Implicit in this is that a secret desire for incest is instinctive, but is so dreadful to contemplate that it is normally repressed, yet smolders subconsciously, resulting in a gnawing sense of guilt. The *Elektra* complex is the same idea but applied to girls.

Now, in *Totem and Taboo*, the title of which encapsulates the big themes of the day in ethnology, Freud laid out a scenario for the origin of matrimony. Once upon a time, the story goes, our ancestors lived in 'primal hordes' (McLennen's term) dominated by alpha males who jealously guarded their harems. This was the concept of ape societies then current, based as it was on scattered anecdotal reports. The turning point was an uprising of the junior males in which the alpha male – the archetypal father – was murdered. In the confused aftermath, a system of marrying was set up to ensure two main benefits: first, each male would have his own female, and second, the rules of marriage would prevent incest.

Among anthropologists, opinions were divided, for it so happens that Freud's theories enjoyed a considerable following among a cadre of them, namely, those of the psychoanalytic persuasion. Others, however, ridiculed it. It was not until 1963, however, that the genius of Claude Levi-Strauss was unleashed upon Freud's theory, in his little book, *Totemism*.[11] By the way, that book also resolved another mystery in anthropology, namely, the explanation of totemism.

Totemism aside, we may mention some of the more obvious points in his critique of Freud's theory, such as the question of why any rules would be needed to prevent incest if we already had an instinctive loathing of it. Indeed, King Oedipus himself obviously had no instinctive knowledge that

Jocasta was his mother, and was horrified only when he learned that she was. This informs us that it is a purely cultural injunction, rather like our horror of eating rats.

3. THE BOASIAN REFRIGERATOR. Franz Boas arrived in American anthropology equipped with a prestigious European education in physics, of all things, and promptly set about the task of transforming the field into a 'real science.' This meant a total house-cleaning. Out with the old poppycock! In with hard facts, real data! He disparaged and forbade any kind of theorizing whatsoever, ridiculing the 19th century speculations, such as those reviewed above, as worthless flights of fancy. Granted, nobody doubts that a house-cleaning was in order, but after a time of Boasian rule, the opinion was increasingly expressed that he had poured out the baby with the bath. For after all, pure facts and data are meaningless if not set into some kind of theoretical framework. Marvin Harris gives the moniker, *historical particularism*, to the Boasian era of anthropology.

Exactly how Boas seized the reins is best left to historians. From my limited perspective, his writings are not spectacular. His specialty was the Amerindians of the Arctic and Northwest Coast. At any rate, beginning *ca.* the 1930s, Boas became the most influential figure in American anthropology, eclipsing even Robert Lowie, and his influence continued through the next generation *via* the many luminaries trained by him. It was Boas who instituted the requirement that every candidate for the PhD perform a year or more of field research, actually living among some previously unstudied people. He saw clearly that primitive societies were rapidly disappearing, and conveyed to his pupils a deep sense of responsibility to record as much as possible about them before they vanished forever. Another big Boasian theme was *language*: every pupil must study linguistics and learn and record the native language, for otherwise, the languages, too, would be forever lost. Still another huge contribution of Boas was his tireless campaigning against racism, the notion that any people on earth were in any way innately inferior to civilized whites. His efforts in this direction, which included repeated testimony before Congress, contributed greatly in this area.

Coming back to the matter of pouring out the baby with the bath, Boas detested any kind of theorizing, most especially about cultural evolution – about how human societies began, how marrying began. But what is the purpose of anthropology if not to draw some kind of picture about how we got to be as we are? Boas ignored and disparaged T.H. Morgan, but it was later shown that Boas had not bothered to read the large body of factual data assembled by Morgan. This was pointed out by Leslie White[12], who rescued Morgan from obscurity, and correctly identified him as the father of American anthropology. Indeed, White was among the first to openly rebel against Boas' injunction against theorizing, by proposing his own theory.[12]

Meanwhile, the purpose of this section has been to highlight the radical discontinuity between the many early theories of matrimony reviewed above, the absence of such theories by later ethnologists. However, the theoretical vacuum has recently been filled by non-anthropologists, considered next – assorted biologists, evolutionary psychologists, sociobiologists, and their ilk.

4. CHILD SUPPORT THEORIES. The leading theories of today are mostly couched in terms of the general theory of biological evolution, or rather, those sub-specialties known as sociobiology and evolutionary psychology. Now, to editorialize slightly, it appears to this writer that these theories are more of the same yarn-spinning that we saw above in 19th century ethnology. Nonetheless, it is now the reigning paradigm, and its mantra is 'fitness'.

That said, the theory of Owen Lovejoy, a sociologist, appears to be widely respected, judging from the frequency with which it is cited in various books and articles. I first encountered it in Johanson's book on the discovery of Lucy, where Johanson narrates Lovejoy's theory at length.[13] Johanson's stamp of approval doubtless had much to do with its ascent to respectability, plus the fact that in the same year, it was set forth in the prestige journal, *Science,*[14] then in *Scientific American.*[15]

His accounts are rather long-winded, almost impenetrable on places, with lengthy digressions into such things as basic equations of population genetics, showing how even a small advantage in reproductive success (fitness) will eventually spread through the genes of the entire population. He clearly subscribes to E.O. Wilson's doctrine of *Sociobiology.*[16] However, these allusions actually have little to do with Lovejoy's theory, except to shroud it in an aura of quasi-scientific mystique. The real essence of it can be summed up very simply: *our original hominid ancestor came down from the trees to walk upright in order to carry food back to the wife and kids.*

How brilliant! At a single stroke, he has resolved not just one but two major riddles of hominid evolution: why our ancestors started walking on two feet, and the origin of matrimony. Well, not exactly matrimony, but *pair-bonding*, being the habit of many animals, especially birds, to take a single mate for life. According to Lovejoy, this offered a major advantage in fitness because, thanks to hubby feeding his mate, she could make more and better babies. Not only that, but pair-bonding gave him his own girl, ending energy-wasteful squabbles about who gets who. It was a win-win deal. But is it science?

We won't bother to explain the details, mainly because there aren't any. Of course, the reader is invited to consult the references and judge for himself. To me, it differs from the 19th century theories only in being phrased in the jargon of the currently accepted paradigm of 'fitness theory', which someone, perhaps Daniel Dennett, has termed the "universal acid" for dissolving all questions. Evolutionary theory was famously criticized by Stephen J. Gould for its habit of writing Kiplingesque *Just So* stories to explain how everything got to be the way it is, and his critiques put a damper on those excesses, but only briefly.

Of course, Lovejoy is not alone in his theory. Here is a snippet from David Pilbeam, the eminent paleontologist, writing in 1970 (but remaining active, at least in 2005) about early hominids:

"Permanent pair-bonds would be the rule ... [because] the offspring have a permanent father to supply food and protection ... the males doing the hunting while females were responsible for gathering vegetable food and taking care of the infants. ... Kinship groups would then increasingly include a biological father." [17]

Questions immediately spring to mind. First, if pair-bonding is such a boon, then why did it not arise in any of the other higher primates? Why is it rare among mammals generally? That is to say, apes have been raising their babies for millions of years without benefit of pair-bonding. The whole ideas is further weakened by the fact that children in many if not most human societies were raised collectively by the group, and food was always shared. Nowhere did the husband selfishly provide only for his own nuclear family. Only recently have anthropologists begun to question the foregone conclusion that pair-bonding (or monogamy) necessarily offers any kind of significant material or economic advantage.[18]

It strikes me that the group-sharing of child rearing, as seen among apes and most primitive human societies, is clearly superior to leaving it to individual couples. Furthermore, we must wonder, what kind of social arrangement could possibly have existed among early hominids to allow individual pair-bonded couples to live harmoniously together. If they lived in isolated nuclear families, they could hardly have defended themselves.

And, what is the special advantage of a "biological father?" Pilbeam's remark on this is dubious in view of evidence that in at least some primitive societies, the role of the father in causing pregnancy was unknown (reported by Malinowski[19,20] but still controversial). More recently, the belief that a child can have multiple biological fathers was found to be "quite common" in preliterate tribal societies.[18]

Even granting some intrinsic advantage to pair-bonding as the initial step towards matrimony, it is not clear why or how that would lead to the institution of formal matrimony. To echo Levi-Strauss, why would formal matrimony be necessary if pair-bonding was already a long-established tradition, or perhaps even an 'instinct?' Likewise, the suggestion that "kinship groups would increasingly include a biological father" seems doubtful because kinship groups are absolutely dependent on matrimony and the assignment of fatherhood, i.e., could never have arisen without prior means of kinship reckoning, which in turn requires matrimony.

5. SEXUAL CHAOS THEORIES. Another popular class of theories is based on the notion that matrimony was necessary to preclude otherwise rampant fighting among males about sexual rights to females. Anthropologist Peter Wilson, for example, writing about the general emergence of modern man, begins by noting that ape sexuality is quite different from human sexuality, and concludes that this must have become a source of conflict and violence:

> "[Sexual] interest and arousal [in apes] is conditioned by the physical changes undergone by the female ... and does not seem to be a source of sensual and emotional feeling. ... [But with humans] if all females were receptive, and all males are ready, then there could be nothing but chaos. ... It therefore seems likely that rules concerning sexual relationships were among the first devised by hominids as a necessary condition for their survival." [21]

It is certainly true that human sexuality has features distinct from apes, as earlier noted, suggesting a period in our evolution during which sexuality took on its present sensuous quality, as a pleasurable end in itself. However, it is hard to imagine this coming about in an atmosphere of violence. Wilson fails to provide a convincingly detailed scenario, leaving us to wonder exactly how it came to pass that, prior to matrimony, human society was a seething cauldron of unbridled lust. One is reminded again of the 19th century speculations. Wilson's scenario seems to be based on observations of sexuality as we know it today, under the law of matrimony, in which all sorts of violence attributable to the sex urge does indeed occur, despite Holy Matrimony. Indeed, to borrow a trick from anthropologist Radcliffe-Brown, one might as well propose that matrimony is the *cause* of illicit desires and sexual chaos, not its solution.

By way of analogy, the European missionaries were shocked by the nakedness of peoples in warm climates, finding them just too erotic to bear, and demanded that they cover their parts, even though nakedness produced no wanton lusts in the natives. By the same token, one could argue that the Victorian ideal of modest attire succeeded only in fanning the flames of desire, to the point that even a bared ankle incited erotic arousal verging on pain. In the same vein, as married couples well know, when sexuality becomes too easy, it loses its luster. So we are left wondering about the particulars of the sexual chaos theory.

Lovejoy's theory also includes elements of the sexual chaos theory. He gets down to brass tacks by inquiring, "What kind of adaptive advantage was conferred on mankind by pair bonding?" Skipping over the fuzzy logic, the theory amounts to another version of the sexual chaos theory. In effect, he argues that a young mother could not very well attend to her kids if she was constantly harassed by horny males breathing down her neck.

> "There are many ways of defusing that [sexual] aggression, he [Lovejoy] explained. That can be done by development of a pair-bonding system. If each male has its own female, its own private gene receptacle, it doesn't have to fight with other males for representation in succeeding generations." [13]

The main objections were already mentioned. It is impossible to debate such a vague thesis. It's like trying to wrestle with a fog or a ghost. As

Feynman, the physics genius, famously said, "It's so bad that it's not even wrong!"

6. WHITE'S THEORY. Leslie White is best remembered today for his theory of cultural evolution[12], subsequently developed by his students[22] and others.[23,24] In essence, White measures cultural evolution in terms of advances in technology – in terms of energy efficiency in exploiting resources – and predicts a succession of types of social organization, from egalitarian bands to chiefdoms to complex stratified societies (states). The general idea is nicely illustrated in Peter Farb's survey of the Amerindians, where he classifies them by White's scheme.[7] However, its predictions are not always observed. For example, the theory predicts that in the presence of abundant resources and high populations, egalitarianism will break down and be replaced by a chiefdom, kingship, or stratified aristocracy, but this did not happen in the rich and populous American Northwest Coast[25] or in Australia,[26] despite ample time.

White is interesting in other ways, too, which is why I so admired him in my early studies. For one thing, he championed the concept of his teacher, Kroeber, that human cultures are *Superorganisms*, in much the way that a whole person cannot be understood just by studying the cells that make up his body. White called this approach *Culturology*, the idea being that cultures work according to laws of their own, which cannot be understood or derived from the psychology of individuals.[27] Kroeber had been severely censured for teaching this conviction, even forced to publicly recant it, so you might think he would have been grateful to White. But he didn't like White's theory. Nor was Marvin Harris, that arch materialist, more than lukewarm towards White's theory, even though it was materialistic *par excellence*.

Now, in writing his book about cultural evolution, White was more or less obliged to begin at the beginning, so to speak, meaning that he had to formulate some idea of how the matrimonial family began. He starts his chapter 4, entitled *The Transition from Anthropoid Society to Human Society*, by reviewing the early theories, as we have, mentioning also the theories of Charles Darwin,[28] and of Robert Briffault in *The Mothers* (1927). He then points out that a good degree of language proficiency was a necessary prerequisite for keeping track of kin. After that, he rambles onto shakier ground, becoming a bit vague, but the following sound-bites convey the spirit of his rationality:

> "A cooperative organization for the purpose of making life more secure was formed in this way: (1) The social aspect of intrafamily relationships was ... made significant in its own right. (2) This was made possible by language. (3) What was done first within the family was next undertaken outside the family ... by means of kinship terms. ... Thus a cooperative organization of relatives was formed ... It consisted of the husband, the wife, and their children, together with the parents of husband and wife ... their sisters and brothers ... and so on." [12]

The purpose of all of this was "mutual aid functions. Each one had duties toward the others ... and in turn enjoyed privileges. They helped each other in the food quest ... in providing shelter ... defense. It was, in fact, a mutual-aid society." However, he side-steps directly confronting the question of how or why matrimony began. He seems to imply that matrimony already existed, before the rise of kinship terms, but he is not clear on this – probably deliberately.

He offers no detailed scenario. Granted, humans cooperate better than apes, but is this attributable to kinship / matrimony? White does not explain how or why the naming of kin would in itself lead to greater cooperation or social cohesion. Troupes of apes or packs of wolves are quite cohesive.

For the rest of his chapter, he is on firmer ground, moving on to consider the importance of kinship in known societies, a field in which he is expert. Along the way, he refutes the notion that matrimony arose to avoid incest. (That theory remains popular with non-anthropologists.) In sum, White's theory for the origin of the matrimonial family is vague and diffuse, faring no better than the others.

7. SUMMING UP. As earlier mentioned, lest it be thought that this chapter is unfairly negative towards the theories reviewed, the reader is at liberty to consult the sources and decide for himself. None of them convinces me.

It may be noted that most of the modern theories are of the Practical Pig variety, i.e., appeal to fitness theory, food provision, avoidance of conflict, etc. This is consistent with the fabric of contemporary thinking in the soft-sciences such as evolution theory, wherein everything happens because of this or that material advantage.

Mercifully, however, at least some eminent anthropologists seem to have escaped that particular sociocultural pathology, notably Claude Levi-Strauss, who holds that people do as they do for reasons of *intellectual* satisfaction, summed up by his oft-quoted remark, that a particular plant has a certain name "not because it is good to eat, but because it is good to think." More about him later.

4.

Interlude: Kinfolk, Clans & Culture

Before moving on, it is quite essential to provide some additional background, hope-fully interesting in itself to those unfamiliar with this topic. The general idea is to provide some perspective on marrying arrangements in the most primitive types of societies, because in the next chapter, we shall discover exactly how and why these arrangements came into being – clans, in particular.

1. CLANS AND KIN. For present purposes, a *clan* is a group of people within a larger society whose members must marry outside of their own clan, into another. In other words, a clan is an *exogamous* group.

We hasten to add that this statement is a somewhat over-simplified, since not all clans are exogamous, and even those which are – or say they are – do not always follow their own rules (see below). Furthermore, some anthropologists now eschew the term, "clan," but the alternative is a forbid-ding set of technical terms devised by various authors for many special cases. Therefore, we will make do with *clan* as defined above: an exogamous subgroup

The simplest case is a two-clan system (properly called *moieties*). If you are a male and belong to clan A, your wife must be drawn from clan B.

After marriage you will both take up residence in either the wife's clan (matrilocal) or your own (patrilocal residence). Both exist, but whichever it is, it is consistently followed in a given society. Similarly, your children will be assigned either to clan A or clan B, depending on the rule of descent, either matrilineal or patrilineal – or in some cases, complicated admixtures, here ignored for simplicity. Our society today, although having no clans, is patrilineal, meaning that the child takes the family name of the father.

All the boys and girls in your own clan are considered brothers and sisters, and sexual relations with them is forbidden as incestuous, even if only remotely related to you by blood. Indeed, eligible spouses in another clan may be more closely related by blood then many of your fellow clan's people. As a matter of fact, marriage with certain cousins was actually *preferred* in numerous societies worldwide, for reasons given later. Marriage within the clan is not only against the rules and forbidden, it is practically unthinkable. There is simply no sexual appeal between members of the same clan, no more than between siblings in our society. But, in the rare instance of a sexual liaison between members of the same clan, punishment was swift and severe. Nearness in blood had nothing to do with it.

Now, what is the purpose or function of these clan-type arrangements, so common in the primitive world? That question will engage our interest later on, but certainly a very prominent consequence is *the creation of networks of kin relations.* Here is Lesli White:

> "Each individual will be personally connected with a number, per-haps all, of the clans of his tribe by genealogical ties known to him. Thus his father belongs to clan A, his mother to clan B. His parents' parents might involve two others clans, C and D.... He would be related to clans also through marriage: his own, those of his siblings, his uncles and aunts, his children, and so on.... In some kinship systems, any male member of my father's clan is my 'father' and his children will be my 'brothers' and 'sisters.' Female members of another clan would be my 'aunts,' 'sisters,' or 'moth-ers.' ..."p. 124.[1]

At least some kinship ties often extended to neighboring tribes. The importance of such ties is well reflected in an account by Radcliffe-Brown of a journey in Australia with a native guide. He observed that his guide felt it was of utmost importance to establish some kind of kinship ties with every alien camp they encountered, for only then could they enjoy hospitality and amicable talk. On one occasion, no such relationship could be found. As quoted by White:

"'That night,' says Radcliffe-Brown, 'my boy refused to sleep in the native camp, as was his usual custom and on talking to him I found that he was frightened. These men were not his relatives and they were therefore his enemies. This represents the real feelings of the natives on the matter.... If he is my enemy I shall take the first opportunity of killing him for fear he will kill me,'" p.121.[1]

Peter Farb tells much the same of Amerindians when they journeyed far from home territory[2] and many others have reported similarly world-wide.

Each clan typically has its sacred *totem*, usually an animal but equally often a star, plant, rock, or almost anything. When it is an animal, it is taboo to kill or eat it, but whatever the totem, it was an object of great reverence. Totemism was long considered a deep mystery of the "primitive mind," and hundreds of scholarly books and papers were written trying to explain it, or to psychoanalyze it. A much more prosaic explanation was given by Claude Levi-Strauss, now generally accepted. He gave the example of a modern military unit nick-named the *Rainbow Company*. The men of the company soon became fascinated by rainbows, even superstitious about them, often wearing rainbow charms or bracelets, in effect adopting the rainbow as their 'totem.'[3] In that view, a clan's totem is nothing more or less than its name, or emblem, which over the centuries took on a special mystique, not wholly unlike the mascot of a modern football team, e.g., the Miami Dolphins. His explanation of totemism was part of his answer to Freud's elaborate but fanciful psychoanalytic theories on such matters.[4]

Each clan in the tribe also has its own important duties, special privileges, role in ceremonies, and elements of prestige. Perhaps for that reason, clans are frequently rivalrous, and tensions between them can be severe. Here is Frank Waters speaking of the Hopi in recent times:

"The clan is still the heart of Hopi society ... To it can be attributed all the virtues of group cohesiveness and loyalty to tradition that mark the individual. Upon it also can be blamed all the evils that still restrict the Hopi within a social mold too narrow and frigid to expand with the ever-widening pattern of modern life," p.148.[5]

Yet, open warfare between clans was exceedingly rare, even practically impossible, if only because everyone had many close relatives in the other clans. On the other hand, even in our own society, murderous crimes against families of in-laws are not rare.

2. DOUBTING THE RULES. It has been known for a long time that native informants have a tendency to paint an idealized picture of their marrying customs, often rather different from their actual practices. Doubtless we ourselves would do the same if asked, for example, to describe our legal justice system to an outsider. However, two speakers at a conference carried this to a radical extreme, Hiatt doubting if clan-based marrying rules had any relevance whatsoever to the actual marrying practices of his Australian study group,[6] and Meggitt claiming that the whole idea of marriage regulation by complex rules of clan affiliation was a myth concocted by anthropologists.[7]

Their claims were all the more audacious – by design? – because the conference was held in honor of Claude Levi-Strauss, who was present, and who by all accounts had made the most brilliant analysis ever of the clan-based rules of marriage, including Australian. They were implying, in effect, that everything he had written on the subject was a high-hatted myth. His reply to their "poison darts" (on p210 and in his concluding essay, p349) strikes me as demolishing their arguments, first by pointing out that their interviews were conducted entirely in a government reservation with informants who had never themselves lived in the old ways, and so reflected mere tatters of a way of life long since decimated. In fact, declining populations often made it impossible to find a spouse in accord with the old rules. Second, he reeled off a list of the early observers whose field studies were conducted well before the system collapsed, namely, those who gave us the historical record, and all of whom corroborated one another's independent reports.

"Should we burn all their books?" asks Levi-Strauss. Under such withering fire, Hiatt backed down, admitting that in his observations, "roughly 90% of their marriages are in accordance with the rules." Enough said.

Nevertheless, to the limited extent that the arguments of Hiatt and of Meggitt might have some merit, they raise an important issue that will interest us later on. In particular, they make it clear that the clan-type divisions in Australia do not exist *solely* for the purpose of regulating matrimony and kinship, and I am sure that Levi-Strauss would agree. Various other functions of the clan systems are often discussed.[8] In the opinion of several scholars, they also function as *intellectual structures* reflecting the native's cosmology, which requires that humans organize themselves in harmonious accord with the larger cosmos. Levi-Strauss himself has pointed out that Australians were highly cerebral, intellectual, and philosophically inclined when discussing such matters.

"Few peoples equal the Australians in their taste for erudition and speculation and what sometimes looks like intellectual dandyism, odd as this expression may appear when it is applied to people with so rudimentary a level of material life. But lest there be any mistake about it: these shaggy and corpulent savages whose physical resemblance to adipose bureaucrats or veterans of the Empire makes their nudity yet more incongruous, these meticulous adepts in practices which seem to us to display an infantile perversity – manipulation and handling of the genitals, tortures, the industrious use of their own blood and their own excretions and secretions … – were, in various respects, real snobs. They have been referred to as such by a specialist born and brought up among them [T.G.H. Strehlow]…."[9]

3. KINSHIP SYSTEMS. The great compendium of kinship terms collected by T.H. Morgan was introduced in the previous chapter. He found that all the systems world-wide could be classified into just five or six systems of kinship terminology.[10] He also discovered close relations between marrying customs, kinship terms, and other features of social organization, and his description of the egalitarian clan and observations on the rise of stratified state-organized societies have also stood the test of time.

Morgan's book of 1870 has been widely praised as "monumental" and other superlatives by such luminaries such as Robert Lowie, G.P. Murdock, Sol Tax, Radcliffe-Brown, Evans-Pritchard, and Leslie White.[11] However, as earlier noted, Boas, ignored Morgan, probably because of Morgan's later attempt to devise a theory of matrimony and cultural evolution. But there may have been another reason why Boas ignored Morgan's kinship work. As Fred Eggan observed, many students find the study of kinship to be exceedingly dull and tedious. I was no exception. The first course I took in anthropology, at the tender age of 18, was taught by a specialist in certain Amerindian tribes, who focused relentlessly on kinship terminologies and clan-type marrying systems, and used Lowie's *Primitive Society* as a textbook. I was very disappointed. This was anthropology? Only in later years did I discover that kinship systems are a lot like mathematics: you have to really dig into it, working problems for yourself, to discover how fascinating it really is – and how it can be applied to solve real-world problems (next chapter). Here is anthropologist Powdermaker:

> "As a graduate student I had taken a year's course in kinship and had become bored with the intricate classificatory systems written on the blackboard and discussed for many hours. Intellectually, I had known their significance, but it was not until I saw one actually functioning, in Lesu, that I understood how it was *the very basis of tribal social structure*. The explanation of why so-and-so is doing this or that, or refraining, was almost always in terms of kinship" p.78.[12] (Italics added, LH.)

Morgan's classification has withstood the test of time, although alternative classifications have since emerged. Leslie White's book gives a good introduction[1] – including on the hidden complexities – but the book-length treatment by Levi-Strauss is widely considered a masterpiece,[13] and Murdock's is also highly regarded.[14] Incidentally, it is interesting to see our own kinship terms, called *Yankee*, analyzed in similar fashion.[15,16]

Now let us correct some over-simplifications we made about exogamous clan-type marrying rules. In Australia, for example, simple two-clan systems are rare compared to more complex divisions into four or eight sections, with complicated rules of marrying that may involve alternating generations, sometimes diagrammed by anthropologists as cycles of wheels within counter-rotating wheels. In addition to clans, there are other kinds of social groupings which may cut across or "intersect" all of the above, and which also figure into the marrying arrangements. Little wonder, therefore, that it was often impossible to arrange real-world marriages in strict accord with the ideal rules. Among other things, the few hundred tribes of Australia each had only a few hundred members, on average, making it almost impossible to have significant numbers of people in each of the clans and other special subgroups.

4. WHO IS MOST PRIMITIVE? Clan-type societies are found world-wide, but in Australia, exclusively that type is found. This means that clans are at least 60,000 years old. It is widely accepted that pure hunter-gatherers are the most primitive type of society, and about three dozen of these – from Africa, Australia, the Arctic, India and some islands – were singled out for discussion at a symposium.[17]

How might one measure degree of "primitiveness?" Indeed, what does the word mean? In the context of this book, it means *closeness to the original marrying system.* The orthodox criterion, however, is way of life. Pure hunter-gatherers do not keep domestic livestock, do not maintain gardens or practice agriculture, do not make pottery, and rarely have permanent villages. Most have very little personal property beyond what can be easily carried: a few weapons, some treasured objects. Another key distinctive feature of such societies is that they were *egalitarian* – no chiefs or kings, or any kind of stratified aristocracy.

Their daily lives were simple and leisurely, as nicely recounted by Turnbull for the case of some Pygmies of central Africa,[18] and by Hart and Pilling for an Australian tribe.[19] Numerous other good books tell of the African Bushmen (e.g., *The Harmless People*, a best-seller), Eskimos (a.k.a. Inuit), and others. Turnbull and Pilling also contributed to the above cited symposium,[19,20] where in several discussions another consensus on hunter-gatherers was reached: most were remarkably peaceable.

Australia, of special interest to us because of its early habitation and comparative isolation, had some 300 tribes (and nearly as many languages), each totaling from a few hundred to a few thousand people. The average population density was low, less than 1 person per square mile, but varying regionally according to resources. The foraging units were typically groups of 10 to 30 people, rather fluid in composition, but they would congregate together periodically for funerals, weddings, celebrations, rituals, initiations, feasting, gossip, and so forth.

In nearly all such tribes, food was ample and required no more than a few hours each day, but there were exceptions such as the Eskimos, who regularly faced starvation.[21] Many hunter-gatherers slept in the open on the ground. In one South American tribe, efforts to vaccinate them soon after first contact proved difficult because their skin was leather-tough.

In Hart and Pilling's account, a major interest of the people he traveled with was *marriage*, for it so happens that among the Tiwi, as in some other Australian tribes, a measure of prestige or 'wealth' was the number of wives a man had. However, the system was set up in such a way that only elderly men could possibly accumulate many wives, usually a mix of pubescent girls and old women. It was not uncommon for men to be still unwed at age 30 or 35. (On the other hand, there was plenty of hanky-panky out in the bush.) This so-called "gerontocracy" was widespread in Australia, and reflects an ancient tradition of respect for elders.

There is no reason to believe that the outlines of Australian society had changed much in the 60,000 years of its habitation. Of course, almost any general statement in anthropology will arouse dispute, and one such statement is this, that primitive societies tend to be fixed and unchanging. Obviously, there will be changes – upheavals from time to time – such as by wars or famine, causing the disappearance of some tribes and the rise of new ones. But certainly, the broad outline of the clan-type of marrying system found in Australia was not invented there, for similar systems are found all over the world, and where they are not, surviving traces of them can often be detected, as we shall see.

Sexual habits apart from marriage are quite variable. Pre-marital sex was widely tolerated, but unwed mothers were extremely rare – possibly because girls were commonly married by the age or puberty or even before, often by prior arrangement. In many societies, considerable sexual license was granted, or even expected, on certain special occasions. Otherwise, however, adultery was everywhere a serious matter. For example, Eskimo men often willingly shared the favors of their wives with other men, but more often than not, the participants preferred to have their trysts illicitly, in secret. Almost always, the word got out, and murder for reason of adultery was said to be a leading cause of death of men. Sometimes years would elapse before the revenge was exacted. There was no penalty for the murder. It was considered justice.

Readable and informative accounts of life in several primitive societies are given in Margaret Mead's best-sellers, such as *Sex and Temperament*[22] (but her study subjects in New Guinea were not hunter-gatherers). Her reporting was later attacked as highly biased and subjective, notably by Derek Freeman.[23] Freeman's arguments were subsequently demolished.[24,25] I personally witnessed a young graduate student jumping to his feet in a lecture hall, loudly impugning her methods and conclusions, but she handled it with graceful aplomb, leaving him blushing.

On the other hand, the attacks on Mead are symptomatic of an important basic issue: almost all anthropological reporting, not just Mead's, has been the target of similar accusations – of biased or selective reporting, or seeking to support preconceived notions. Indeed, precisely such fractious disputes have been at the heart of the sometime rancorous divisions of anthropology into its many "schools" and shifting fashions, our next topic.

5. ANTHROPOLOGY AND ME. Some readers will notice that many of the key references in this book are prior to 1980. This should be expected from the Preface, which candidly explains that this book was first drafted way back then. Does that mean that it is outdated, no longer valid?

No. In the first place, primitive societies are a lot like real estate: "they ain't makin' any more!" That is to say, by roughly 1960, virtually all primitive societies – the meat-and-potatoes of anthropology – were essentially extinct, or so hopelessly corrupted by contact with westerners as to be useless for gaining insight into their pre-Columbian ways. As a result, the old accounts are *the only* accounts. This has caused a bit of a crisis in the field, and so it behooves us to briefly review some of the history of it.

First, let us sing some praise for the many great men and women who gathered what precious knowledge we have about primitive societies, at low or no pay, and at risk to life-and-limb, in the four corners of the world. More than a few never made it back, or returned weak with tropical diseases. They knew full well that their subjects were rapidly disappearing and took it as their sacred duty, with as much zeal as any Christian missionary, to collect what knowledge they could, before it was too late. The closest we have to such unsung heroes today are those now racing to protect, conserve, and record the world's vanishing species of plants and animals. The immense variety of primitive societies is our only window on the customs of our ancestors, now all but obliterated by the steady homogenization of the modern world.

We already touched on the rise of American anthropology in Ch. 3, at least to the era of Franz Boas. His effort to make a "real science" out of anthropology by collecting "just the facts" was part of a historical trend, for we see exactly the same approach taken by Watson and Skinner in the field of psychology, who founded behaviorism, as discussed elsewhere.[26] After Boas and other early American luminaries such as Robert Lowie, the next big thing in American anthropology was Alfred Kroeber, mentioned earlier.

Fortunately, however, other countries had their own brands of anthropology, notably two, English and French. England had its own 'father of anthropology', in the person of E.B. Tylor, of interest later, followed by other luminaries such as Malinowski, Radcliffe-Brown, and others. French anthropology was founded by Rousseau, followed by Durkheim, Mauss and others, capped off by Levi-Strauss. Thus American, British, and French anthropology were spoken of as distinct schools. Cutting across all of these were other 'schools' or approaches, such as those based on psychology – Freudian or other – with spin-off movements such as *Cognitive Anthropology*.[27] At the same time, there arose what I like to call the Bean Counting school, actually several, with the aim of being more objective and "scientific" by trying to explain cultural features in terms of material resources, leading them to measure crop yields, water usage and distribution, energy expenditures, and so forth – including bean counting, literally. Results were often quite interesting.

For those who might be interested, Marvin Harris surveyed all of these movements and more, as of 1968[28] (later updated). He draws an overarching distinction between Materialist *vs.* Idealist approaches to understanding culture, the former in terms of resource management, the latter in terms of psychological motives or principles. He himself is strongly of the materialist persuasion, taking his founding heroes to be Herbert Spencer and Karl Marx (the latter popular in left-leaning academia at that time), therefore his book is strongly biased in that direction, at the considerable expense of, for example, French structuralism and Levi-Strauss. However, it is a very readable and scholarly account of anthropology up to then.

From all of this, one may rightly conclude that the meaning of 'cultural anthropology' depended very much on which college or university you studied the subject, as each was dominated by one or another approach. Now, this fractious divisiveness among dynasties was not altogether a bad thing, because at least the anthropologists were openly admitting that they had no clear idea of what they were doing, in contrast to other fields (such as psychology, sociology, and evolutionary biology) that present a united front to the world, as if they had everything sewn up. What a laugh.

The confusion of the field began to dawn, perhaps in the 1970s, with article titles such as, "Has anthropology betrayed its mission?" [29] Discussions were common about whether or not anthropology should even be considered as a "science" at all. Perhaps it was better thought of as a field of *history*, for after all, historians also record peoples and events, and then try to formulate theories about causes. Then again, perhaps it was largely a field of *literature* – or akin to literary criticism? – since nearly all of the big luminaries were masters of exposition, often at their best when assailing one another. One of the few nice things about getting old is looking back on all this sturm-and-drang and seeing patterns, such as the writing style of the likes of Lesli White and Marvin Harris in the context of others of the era, such as Bertrand Russell, H.L. Menken, or B.F. Skinner, all so acerbic, witty, clever – and fractious!

That brings us up to the modern era, which I know even less about, but the new regime seems clearly to belong to Clifford Geertz – another gifted writer, by the way. A typical accolade:

"We should be grateful that Clifford Geertz chose anthropology. Described as the single most influential anthropologist of recent times, he resisted reductionism and chose meaning-centered analysis." Maggie McDonald, in brief book review[30].

The exact meaning of "meaning-centered analysis" is not crystal-clear, but in the course of reading some of his books, one thing does become clear, that he is a very good anthropologist – and writer! But don't expect much in the way of scintillating theorizing. We shall continue this thumbnail exposition later on, when we shall have occasion to speak further of Claude Levi-Strauss, who, by the way, appears to have exerted a powerful influence on Geertz.

6. NATURE, CULTURE, & MATRIMONY. A long-standing question in cultural anthropology has been how to define *culture*. This might seem like a trivial and purely academic matter, but it turns out to be of central importance, with real bearing on our thesis, as will emerge. Kroeber supplied the definitive survey of concepts and definitions of *culture* as of 1952.[31] (Incidentally, Geertz worked on that opus, apparently then a student). However, no consensus was reached in that book, and other efforts came on its heels, e.g. Cafagna's.[32]

Very briefly, there are just a few broad concepts, perhaps only two, or three. One is that culture is uniquely human and consists of the total lore transmitted from generation to generation, independent of genetics – or, in a word, is the totality of the *traditions* of a people: their language, religion, marrying customs, hunting practices, pot-making, and so on. This concept often took the form of long lists of items in textbooks, commonly beginning with universals such as fire, tools, kin relations, art, music, etc. One problem with this definition is growing evidence that animals exhibit many of the same features, in which case "culture" under this definition is *not uniquely human.* Even birds sing and dance, make tools, have language dialects, transmit learned behavior, etc. However, the main debate has centered on apes.[33-46]

Quite different is the French tradition of anthropology, holding that the biblical fall of man was a real historical event, marking the transition of man from an animal state of *nature* to the human state of *culture*. This position stems from J.J. Rousseau, as in the following from Levi-Strauss:

> "In almost modern terms, Rousseau poses the central problem of anthropology, *vis.*, the passage from nature to culture. More prudently than Bergson, he abstains from introducing the idea of instinct, which, belonging as it does to the order of nature, could not enable him to go beyond nature. ... For Rousseau, moreover, affective [emotional] life and intellectual life are opposed in the same way as nature and culture. ... *The advent of culture thus coincides with the birth of the intellect.*" [3] (Italics added, L.H.)

It is not extraneous to remind the reader of our modern fascination with the idea of being "natural," whatever the word may mean, and to point out that this idea is meaningful only in juxtaposition to "cultural." Indeed, the popular meaning of "being cultured" refers to high-brow literature, music, opera, and the arts, all of which are quite opposite to things "natural."

Now, it just so happens that a number of anthropologists have discovered that primitive peoples, too, despite the popular illusion of them being "natural," are acutely aware of this distinction, and are constantly seeking to balance themselves against excesses in either direction, as if fearing a slide back to the imagined chaos of animal nature, while at the same time vigorously resisting and opposing excesses of things cultural, i.e., things imagined to be of human invention. This is a recurring theme in, for example, Levi-Strauss' 5-volumes of myth analyses.

These opposites are best illustrated in our own society. Excessively cultured people are those who are overly concerned with money, laws, booklearning, proper speech, etiquette, and so forth, and are commonly scorned as prissy, squares, geeks, nerds, or fags. On the other hand, those of a violent or criminal disposition, or who speak poorly, or are rude, careless with money, illiterate, or hyper-sexual, are disparaged as Neanderthals, goons, and other pejoratives, for they are excessively natural. The ideal man, of course, is a perfect admixture of both. James Bond, for instance, though a man of few words, tough as nails, and endlessly sexual, is nevertheless Oxford-educated, perfectly mannered, and seems to know everything cultural. The balancing act is everywhere apparent. Prostitutes, for instance, counter-poise their excessive sexuality (too natural!) with excessive trappings of culture: heavy cosmetics, spike heels, and fancy clothes, not to mention selling love for money. Indeed, certain social groups are identified, if only in the popular imagination, as representing these two extremes: Jews are viewed as excessively cultured (bookish, legalistic, money-centered) while Afro-Americans are seen as "too natural" (ill-spoken, little respect for money or laws, etc.). Either extreme is despised, but at the same time covertly envied by those who judge themselves deficient in one or the other quality.

And now, what exactly was the historical transition from animal to human, from *nature* to *culture*? What was this "fall of man" from grace? We shall see that it was nothing more or less than the invention of matrimony. Nothing is more essentially cultural, or "less natural," than getting married. Furthermore, it was clearly recognized even by some of the early ethnologists, such as L.H. Morgan, H.S. Maine, that the specific marrying customs and associated kinship ties are absolutely central to the social fabric of any society. But, how, and when, and why did it start? Read on.

5.

Tylor's Conjecture

1. THE PRE-MATRIMONIAL SCENE. We can never know the details of human life before the institution of matrimony, but a few things can be said with reasonable confidence. Most will agree that in outward appearances, it differed little from known hunter-gatherers, which is to say, they lived in small bands of a few dozen individuals. Larger groups cannot be supported by that lifestyle. We may also assume that they were egalitarian – no chiefs, no alpha-males – for reason earlier given: weapons were the great equalizer. Of course, this does not mean that everyone was equal. Respected leaders, usually elders, outstanding hunters, healers, shamans, warriors, etc., always emerged, and many societies were marked by incessant struggles to gain prestige, often by accumulating "wealth" of some sort, or by feats of bravery, e.g. daring to taunt or steal from a neighboring band.

A big question is, of course, the sexual habits of the pre-matrimonial people. Details are unknown but in the absence of matrimony, there could have been only one possibility: limited promiscuity. The role of paternity was probably unknown, since we earlier noted that some historical societies were apparently unaware of the father's role in making babies (chapter 3.4). Of course, siblings knew their relationship to each other and to their mothers – as do chimpanzees – and perhaps were aware of their mother's siblings and their children, but beyond that, kin reckoning would have been impossible, and was probably not a central feature of social organization. There is every reason to believe that within the band, relationships were warm and compassionate, which does not preclude a degree of bullying and a pecking-order, perhaps especially with regard to sexuality.

Relations between neighboring groups was probably similar to those seen in historical primitive societies: strained, antagonistic, and mutually wary, yet marked by regular polite contacts between their respective elders, Overt hostilities flared up occasionally (or regularly) but in most cases these were ritualistic skirmishes or tit-for-tat sorties, rarely large-scale mayhem. In many societies, such hostilities had gone on as long as anyone could remember, probably for millennia, resulting in a "perpetual state of high-anxiety," as one observer described it. At the same time, however, mutually antagonistic groups almost always maintained contacts, such as for purposes of trade and bride exchange. The situation between pre-matrimonial bands may have resembled that seen, for example, in the central highlands of New Guinea:

> "[Warfare] is an accepted social institution. The actual result could be several deaths ... destruction and burning of a village ... abduction of women ... However, wholesale destruction is the exception rather than the rule. ... There is a repetitive quality about this, interspersed with peace ceremonies and other inter-district gatherings. ... Thus Jate, Usurufa, Ofafina, and Fore are bitter enemies of Kogu, but they are also its main source of wives as well as its guests for certain festivals, ceremonies, and so on." [1]

Were prematrimonial bands yet allied into what we might call "tribes?" Possibly, but only for brief periods, since the crucial basis for such affiliations – kinship networks – could not have existed prior to matrimony. In fact, we shall see that one specific kind of such alliance was precisely the origin of matrimony.

2. THE TYLOR CLUE. Very possibly the most frequently quoted passage in all of anthropology is the following, from E.B. Tylor, known as the father of British anthropology:

"Among tribes of low culture there is but one means known of keeping up permanent alliance, and that means is intermarriage.... Exogamy [by] enabling a growing tribe to keep itself compact by constant unions between its spreading clans, enables it to over-match any number of small intermarrying groups, isolated and helpless. Again and again in the world's history, savage tribes must have had plainly before their minds the simple practical alternative between marrying out and being killed out." [2]

The basic idea here is that exchange of brides – "marrying out" (exogamy) – could be an effective way of forging a bond between antagonistic neighbors, certainly better than being "killed out." Tylor did not propose this idea to explain the origin of marrying, which he believed had already existed, but only the origin of *exogamy,* a.k.a. clans, or the "incest taboo." Neither has any of the many who have quoted this passage suggested anything beyond Tylor's claim, that bride exchange could forge a bond between otherwise antagonistic bands. For example, Leslie White states that "incest prohibition and exogamy preceded clan organization ... exogamy being a prerequisite for clan structure"[3] (p88).

We shall find otherwise. In fact, the solution is so obvious from this clue that I cannot fathom why none of the many authors who have quoted this passage failed to see that with but the slightest modification, the same concept could account for the origin of matrimony itself.

That is to say, as Levi-Strauss has repeatedly shown, matrimony can hardly be separated from the exogamy rule, for its very essence seems to be bride exchange, making it all but impossible to imagine how marriage could have arisen prior to the exogamy rule. Furthermore, we have seen in the previous chapter how implausible it is to try to explain the origin of marrying independent of exogamy and kinship tracking. Accordingly, a really satisfactory theory must explain the rise of *all three* of these features simultaneously: marriage, exogamy, and kinship tracking. Let us call it the MEK complex.

In further support of this contention, it has been pointed out by many authors that the kind of alliance that Tylor spoke of, based on bride exchange, has never been actually observed to happen, even though we would expect to see it quite often if it really was an effective means of settling disputes and forging alliances. Indeed, this absence of observed examples has led many to dismiss Tylor's idea entirely, and to cite it only for its historical interest.

Not only has bride exchange not been observed as a means of settling disputes or forging alliances, but it is impossible to imagine that happening at all: who would willingly deliver a beloved daughter into the hands of an enemy? Another reason why it has not been observed in historical times is that all known societies already possessed internal exogamous marrying, which is to say that Tylor's hypothesis makes sense only if applied to pre-matrimonial conditions.

3. A PLAUSIBLE SCENARIO. Let us now climb into our time machine and journey back some 70 thousand years to do a little fieldwork. Suppose we have identified a number of hunting bands dotting the study region, each numbering a few dozen individuals. As expected, they are pre-matrimonial and sexuality is loosely promiscuous, although specific habits vary from band to band. Occasionally, juveniles of both sexes depart their home band and join a neighboring one. Relations between neighbors are often strained but serious hostilities are rare or sporadic.

Bearing in mind that we are speaking of humans already in possession of language, it is not unexpected that we observe neighboring bands occasionally meeting and speaking near their territorial boundaries, exchanging news and civilities through their elders, the others standing back, silently appraising one another. We are fortunate to witness the trade of some items such as flint, seashells from the coast, dyes such as red ochre, beads and whatnot.

Despite the mutual wariness, a strong element of curiosity about one another, particularly about members of the opposite sex, is evident by whispers and giggles among the youngsters. Such interest is not difficult to fathom. Although promiscuity might seem pretty racy, when you imagine it confined to a group of some 35 people living intimately together, half of whom are either elderly or little children, it could easily become rather hum-drum as the years wore on. As they say, familiarity breeds contempt.

Next in our field-notes is the observation of instances of temporary cooperation between neighboring bands, occasioned by the seasonal return of herds of game animals, for the purpose of large-scale hunting, aided perhaps by dogs and setting fires. The hunt is a huge success, culminating in a large bonfire, prayers of thanks, pledges of undying allegiance between the bands and – heavily underlined in our notes – a goodly measure of sexual hanky-panky between the two groups, marred by shouts of anger.

It is common knowledge that we often find that strangers or foreigners exert a stronger sexual attraction than the local neighborhood kids. If this needs confirmation, many scholarly studies by psychologists and others have affirmed it to be true, e.g., studies of marriages of residents of the Israeli kibbutzims has shown they prefer to take mates outside of their home group, even though there is no rule or pressure to do so.

Returning to the field, our fellow anthropologist, 50 miles downriver, reports another observation of a two-band alliance, this time to repel a common enemy. They succeed, killing several and sending the rest fleeing, resulting again in a jubilant celebration lasting all night – and more sexual commingling, described by our colleague as an orgy. Fond bonds are formed, though the scene is tarnished by a number of loud disputes between individuals, calmed by their fellows.

Our field-notes agree that in neither case was the temporary alliance and sexual exchange a means of settling a dispute (as Tylor proposed). In fact, it is difficult to imagine sexual exchange in an atmosphere of suspicion or hostility. Such an event would only have occurred between neighboring bands that were more-or-less friendly, such as following a collaborative hunt or victorious fight against a common enemy. The sexual aspect would be merely the consummation of long-standing mutual curiosity, including, of course, sexual curiosity, if not lust.

4. THE MORNING AFTER. In our hypothetical field notes, spanning 400 years of observation of approximately 42 bands in the study region (ranging from 31 to 53 individuals), the observation of an initial sexual commingling was made 73 times. Of special interest is following out the aftermath of those 73 cases. Numerous scenarios are likely and all of them almost certainly actually occurred on multiple occasions. However, only a rare few would have resulted in a permanent two-clan system, replete with exogamous marrying. Here are the possibilities.

(1) Fusion. In some cases, intimate relations between two bands would have led to their total fusion, i.e., formation of a larger band identical in structure to the two that created it, a kind of coalition. However, this outcome would have had no revolutionary consequences since no basic change has occurred. In fact, it had little potential for long-term stability

because the population of hunting-gathering foraging bands is always limited by the resource density. The enlarged band, although breaking into smaller foraging groups, may well have retained its collective affiliation for some time, perhaps decades; but eventually the early euphoria of the camaraderie would have faded or soured, as such things are wont to do, and matters returned to their former state.

(2) Regular trysting. Another outcome repeatedly seen is regular repetition of the original convocation, perhaps at each season hunt or full moon. The groups retain their separate identities. That is, the two groups returned to their camps and continued as before, except for the regular partying. Our field notes show that this was more common than fusion, and it is easy to understand why: the elders and leaders of each were presumably proud of their bands, certain to be more than little resentful or jealous about sharing their influence and prestige. Nevertheless, fueled by sexual interest, or even romantic love affairs, the two groups would periodically meet to rekindle their bond, doubtless an exciting event to anticipate. One can easily imagine the heated conversations and debates in the two camps on each morning after – including the first tearful accounts of romantic love, the smitten boys and girls speaking glowingly of passions more ardent than any they had ever before known.

However, neither would this arrangement have had any real built-in long-term stability or revolutionary consequences, for no permanent weld had been established between the two bands. At any time, some trivial incident or quarrel could touch off hostilities between them, stopping the conjugal visits and destroying the alliance. The fact that some children were fathered by males of the other band would have been of no consequence, since the role of paternity was probably unknown,

(3) Other inconsequential scenarios. A number of other scenarios following the initial commingling are recorded in our field notes but hardly need discussion, since neither did they have any major long-term consequences. For example, groups of young males (or females) sometimes emigrated *en bloc* to another band, either because of disputes back home or in order to live with their new-found darlings; but this was seen to entail resentments in one or both camps, souring the initially auspicious relations. No long-term stable consequences eventuated.

5. ON THE ROAD TO MATRIMONY. Upon reflection, the two main outcomes listed above, fusion and periodic trysting, are opposite extremes. Neither would lead to an enduring and revolutionary social system. Between these extremes of entropic mixing, however, is a third possibility: an agreement that sexual relations could take place *only exogamously.* That is to say, boys of band A could have relations *only* with girls of band B, and *vice versa.* It could well have seemed an appealing arrangement, and a means of resolving certain problems that otherwise bedeviled the two groups.

This agreement was doubtless reached only after much reflection and deliberation by the elders of both A and B, each hearing their constituents' impassioned accounts of loves, broken and new, weighing these tearful entreaties soberly against the larger political and military ramifications.

It is not difficult to imagine why this issue would have surfaced. Indeed, it was almost inevitable. For any such relationship between bands A and B would be certain to result in equal measures of heartbreak, jealousy, and new-found passionate love. It was a delicate matter. But we take it as a fact of history that in at least one such case, the wise decision was made to insist that sexual relations between the two bands be strictly exogamous. This simple decision, which was almost forced upon them, would have far-reaching implications for humanity.

For the majority, this would not have been a bitter pill to swallow, granting that most were pleased with their exciting new consorts, so much better than the often tedious inter-personal relations in their original bands. It was really the only fair way to settle the squabbling. Just about everybody endorsed the idea.

6. WALTZING DOWN THE AISLE. The rest of the story is practically self-evident, and would have played out with a high degree of inevitability. Promiscuity under this new arrangement would have been unworkable. In the original bands, each person was raised in a long-established socio-sexual tradition, but in the new alliance, there was no prior tradition to govern such delicate matters, so that fights and squabbles would have erupted, the younger males of both groups behaving like foxes in a henhouse. In this circumstance, something like "sexual chaos" would indeed have been a serious threat to the stability of the alliance. The only way to preserve it was for the elders to proclaim the law of one-on-one relationships, demanding vows of monogamous fidelity under the law of Holy Matrimony.

The resulting system, which has endured to the present, was really an intellectual triumph, a brilliant bit of socio-political engineering, so to speak. Indeed, one may view it as the first socio-political ideological experiment, or at least, the first to work well enough to endure for 65,000 years.

7. KINSHIP LAW. The final stroke needed to complete the system was *kinship law.* The need for this would have emerged within a few years of establishing the original system of exogamous matrimony, because the question would inevitably have arisen, *who will the offspring of these unions marry?* The now-obvious answer is that each child would have to be assigned membership in one or the other band – now clan – either the mother's or the father's, since otherwise, the rule of exogamy could not be perpetuated to the next generation. Granting that a choice was agreed on, the system is now complete, containing all the rudiments of historically known clan-type systems, for what we now have is two bands (clans) welded together by exogamous matrimony, yet both will perpetually retain their separate identities since all offspring are assigned to one or the other band in equal numbers.

Thus, by a single simple and perfectly plausible scenario, inspired by Tylor's conjecture, we have accounted for the entire MEK complex (marriage, exogamy, kinship), in comparison to which all other efforts to account piecemeal for the origin of these institutions seem hopelessly inadequate and flatly incredible.

8. CROSS-COUSIN CLINCHER. Among the more exciting "Eureka" moments in developing this theory was the realization that it explained the worldwide prevalence of cross-cousin marrying (and distaste for parallel-cousin marrying), even in societies with no remaining trace of clan-type organization. At the time, I believed that this was my own original discovery, and in a way it was, although I later learned that the connection between clans and cross-cousin marrying had long been known.

Leaving that for later, it must be explained that a *cross*-cousin is a child of your father's sister or mother's brother (different genders), while a *parallel* cousin is a child of your father's brother or your mother's sister (same genders). Now, suppose you are a male of clan A and looking for a wife. Are any of your cousins eligible? Assuming that this society is matrilineal, then your mother and her sisters and brothers are also clan A. Thus, your mother's brother must marry a woman of clan B, and their children (your cross-cousins) will be clan B, so those daughters *are eligible* for you to marry. By the same rule, your mother's sister must marry a man of clan B, but their children (your parallel cousins) will be clan A, the mother's, because of the matrilineal assumption, and therefore are *forbidden* for you to marry, since that would amount to incest.

Similarly, consider the cousins from your father's side. Your father comes from clan B (because he married your mother, of clan A) so his sisters and brothers are all B as well. Thus, your father's sister (abbreviate FaBSiB) must take a husband from clan A and their children (your cross-cousins) will be clan B, because the mother is clan B, so those daughters *are eligible* for you to marry. Your father's brother ($Fa_B Br_B$), on the other hand, takes a wife from clan A, so their children (your parallel cousins) belong to clan A; hence, daughters among them are *forbidden* for you to marry. That is, since you are also clan A, it would be incestuous to marry your father's *brother's* daughter ($Fa_B Br_B Da_A$). But it's okay to marry your father's *sister's* daughter ($Fa_B Si_B Da_B$). It's even more than okay; it is commonly *preferred* to do so.

Essentially the same result is obtained if the system is patrilineal rather than matrilineal, except that the symbols are interchanged because now the daughter belongs to the clan of the father, not the mother. Rather than repeat all of the above, we can explain more compactly in table form, using the above symbols and a few more, namely, wife = Wi, husband = Hu, daughter = Da. The self is termed Ego and is again assumed to be male and of clan A.

- $Fa_A Si_A$ + Hu_B yields Da_B, allowed to Ego_A.
- $Fa_A Br_A$ + Wi_B yields Da_A, forbidden to Ego_A.
- $Mo_B Si_B$ + Hu_A yields Da_A, forbidden to Ego_A.
- $Mo_B Br_B$ + Wi_A yields Da_B, allowed to Ego_A.

In every case, whether descent is matrilineal or patrilineal, only the *cross*-cousins are eligible, the *parallel* cousins being always forbidden, amounting to incest. As I see it, the worldwide prevalence of the cross-cousin preference, and avoidance of parallel cousins, in hundreds of societies with no surviving trace of clan organization, is powerful evidence in support of our claim that matrimony originated from the exogamous union of two bands, as described.

9. THE RECEIVED WISDOM. As mentioned above, this realization was a real Eureka moment, since it could hardly be pure coincidence that cross-cousins are so widely preferred in marriage, even in hundreds of societies with no surviving trace of clans. So imagine my dismay when further reading led me to find that a number of experts on kinship and marrying were long aware of this connection.

However, upon closer inspection, it turns out that they all made the same unnecessary assumption that Tylor did: they all assumed that marrying already existed, *before* the advent of clans! Our scenario shows that the alliance giving rise to clans was itself the cause and origin of marrying, and all that goes with it – exogamy and kinship reckoning. The following quotation from Leslie White illustrates:

> "A relationship between dual organization and cross-cousin marriage has been recognized for a long time, but explanations of this connection have varied considerably. Tylor felt that 'cross-cousin marriage ... must be the direct result of the simplest form of exogamy, where a population is divided into two classes or sections.' At one time, Rivers believed that cross-cousin marriage 'has probably arising in most, if not all, cases out of this [dual] form of social organization.' Furthermore, Rivers and others have held that the custom of cross-cousin marriage may have outlived the dual organization that produced it [so that] the mere presence of cross-cousin marriage proves the former existence of dual organization." [3] (p165)

He goes on to point out an objection to this reasoning, noted by several authors, and then to explain how his own theory "shows how cross-cousin marriage could produce a dual organization of society." We have proposed the exact opposite, that the dual organization *came first,* and automatically produced the cross-cousin preference as a persisting feature of a two-clan origin, even long after clan-type organization was abandoned in many societies. That is to say, his theory, like the others, takes it for granted that marriage existed *before* the dual organization (two clans, a.k.a. moieties), whereas we have argued that marriage arose as a *result of* the two-group alliance. The scenario we have sketched is simple, clear, and plausible, even compelling, compared to the tortured logic of the alternatives holding that matrimony came first.

Finally, we must deal with another objection to our hypothesis implied by the reasoning of Tylor, Rivers, and others, explicitly raised by Westermarck, Robert Lowie, Leslie White, and perhaps others. The objection is this. The commonly observed preference for cross-cousin marrying is a preference for the *biological* cross-cousin, whereas in the original two-clan system, all people of the same clan were classified as "brothers" and "sisters," and our concept of "cousins" was entirely independent of biological relationship. Hence, the question arises, why is the *biological* cross-cousin so widely preferred in historical non-clan societies?

The answer is not difficult. Whenever the clear-type system breaks down, or is abandoned, for whatever reasons, what remains of it is the nuclear family and the "family name," which is what we have in our society. In other words, the "family name" is the sole surviving trace of the original clans. For us, the exogamy rule amounts to marrying outside of one's family. Hence, *biological* families are the surviving traces of clans, making it easy to imagine how the cross-cousin marrying preference would be a natural outcome of the disintegration of clan-type structure.

So everything fits, and is so clear and compelling – at least, it is to me, but maybe I'm missing something – that the only remaining mystery in mind is why this explanation has not been written up before. And the only answer to this question that I can think of is the habitual belief that everything in evolution, cultural or biological, needs to happen very slowly, in tiny little steps, and in particular, the prevailing assumption that marrying must be of very great antiquity, followed much later by the exogamy rule and kinship structure. But, as pointed out above, that line of though makes no sense. It just doesn't work. What does work is the scenario we have sketched in this chapter, that the origin of matrimony, and indeed, of the whole MEK complex, was practically an overnight invention.

6.

Out of Africa

When this book was first drafted, back in the 1970's, it made a bold prediction, namely, that all of humanity known in historical times sprang from the original marrying society. It is now known that all humanity did indeed spring from an initially small population. In this chapter, we review that exodus, and consider the role of matrimony in out cultural origins.

1. THE NEW *HOMOS*. Recall that chapter 2 took us up to *Homo erectus*, because of theories that marriage, or at least pair-bonding, goes way back. But the previous chapter suggests a more recent origin for matrimony, so we must pause to fill in the later evolution of our species, *vis a vis* the question, when in time did marrying originate? That is to say, the scenario of the previous chapter excludes an extremely ancient origin. To briefly reiterate, after the long reign of the australopithecines, the first of our genus, *Homo*, appeared, in the form of *Homo habilus*, the "handy man" who made the first known stone tools. Homo and his early tools dates to about 2.3 million years ago (mya).[1] Some recent discoveries suggest one or more intermediate links, but here we are interested only in broad outlines.

Then came *H. erectus*, the first to move out of Africa, showing up at many places in Asia and Europe. One of the oldest specimens outside of Africa was an *H. erectus* with persisting features of *H. habilus*, found in the Republic of Georgia.[2] It was believed that *H. habilus* evolved into *H. erectus* and was displaced by *H. erectus*, but some argue that they coexisted as separate lineages for a half-million years.[3] Others are skeptical.[4] We already mentioned the advances made during the long reign of *H. erectus*, notably, much of the increase in brain size that took place between Lucy (350 cc) and ourselves (1400 cc). It is likely, too, that the reign of *H. erectus* was the "Golden Age" of our lineage, during which most of our fine social, sexual, and intellectual sensibilities took shape.

It is remarkable that after such a long reign of *H. erectus*, our own modern species emerged rather abruptly, roughly 300,000 years ago, named *Homo sapiens* (including Neanderthals, discussed later). The earliest known examples were found in Ethiopia, one dating to 157,000 years (157 kilo years ago, kya),[5] another to 198 kya.[6] These eventually spread out of Africa and replaced *H. erectus* world-wide, in a manner to be reviewed.

Meanwhile, we are curious about exactly how and why *H. sapiens* suddenly arose. The orthodox view is couched in Darwinian terms, i.e., that some sort of climate change, or sudden genetic mutation, suddenly launched Modern Man. In other words, the orthodox explanation for all of evolution is that everything was accidental, either by random mutations or by external forces like climate. But we have been led to a very different view, that species *invented themselves*, by way of novel strategies, discoveries, preferences; and that recent human evolution was no different in this respect. As a matter of fact, one recent study found that two early "bursts of innovative behavior cannot be explained by environmental factors,"[7] nor did a genetic study find any evidence of the expected "classic selective sweeps" marking human advance.[8]

In the case at hand, a very good candidate would be the invention of language, or a major step in that direction. As explained in a planned companion book on the origin of language, it arose (like matrimony) as practically an over-night invention. On the other hand, because of the many major anatomical and neurological changes required for speech, it must have been developing for a very long time before the invention of true human language (THL) as we know it today.

Like many major cultural innovations, THL must have been preceded – *foreshadowed* – by a long period of preliminary development. To illustrate by analogy, modern chemistry was foreshadowed by a long period of chemical tinkering, going back to the stone age, by which we mean the mixing of all sorts of brews and magical potions, usually involving the transforming power of fire. This eventually led to the alchemical tradition, which in turn spawned modern chemistry. From this we surmise that humanity long held a more-or-less instinctive conviction that *it was possible* to mix certain things in just the right way to get a desired result. Indeed, modern chemists can assemble molecules almost as easily as children assemble Tinker-Toys, to make things like plastics, rubber, explosives, drugs, and so on Bronislaw Malinowski reported that his islanders believed that their canoes could even be made to fly, if only they knew the right magic. We now possess that magic. Edgar Allen Poe wrote charmingly of numerous examples of modern "magic," long sought by primitive peoples, as if they had known all along that *it could be done.*

Hence, a major advance in language is a likely candidate for explaining the abrupt rise of *H. Sapiens*, displacing *H. erectus*. Such an advance would have originated in a single small group which thereafter expanded.

The origin of *H. sapiens* is currently a hot topic, fraught with controversies, but much has been learned in recent years. Among the more revolutionary findings is evidence of important advances, previously thought to have arisen much later, in Europe 20-40 kya, actually arose much earlier, back in Africa. For example, shell beads, marking the advent of personal ornamentation and "symbolic thinking" were found in Israel and Algeria, dated to 75 kya and 100 kya.[9] Engraved ostrich egg shells, also said to be clear evidence of "symbolic thinking," were unearthed in South Africa, dated to 60 kya.[10] Evidence of pressure flaking, an improved technique for making very sharp and slender weapon points, was found in South Africa dated to 75 kya, a style previously thought to have originated not more than 20 kya[11,12]. Another South African site yielded evidence of a decorative pigment, red ochre, dating to 65 kya, and indirect evidence of the use of traps and snares for hunting.[13] Several South Africa sites are now yielding evidence of two bursts of advances between 80 and 60 kya, separated by about 7 kya[7], but we can expect revisions in the near future when these sites are explored to their full depths.

2. OUT OF AFRICA. The big news in paleoanthropology of the last 20 years has been the almost complete triumph of the Out-of-Africa hypothesis, here abbreviated OAF. Its rival theory is known as the "multi-regional hypothesis," or MRH for short. The former argued that *all humanity today* stemmed from a very small initial population emerging from Africa, about 60 kya. The latter argued that the world's populations arose independently near their present locations, pointing out, for example, that the *H. erectus* specimen known as Peking Man exhibits features characteristic of present day oriental populations. Indeed, such questions linger, but on the whole. OAF is overwhelmingly supported, largely by the powerful new tools of genetic science.

In the early 1990s, the OAF hypothesis gained considerable support with some studies of mitochondrial DNA (mtDNA) from various peoples around the world, indicating that they all stemmed from a single ancestral mother, dubbed the African Eve.[14] This was an exciting finding, and that method was widely applied, such as by Bryan Sykes, who in his book traced all Europeans to just seven ancestral mothers.[15] The method works, or seems to work, because the mitochondria within our cells (which produce our energy and were once free living bacteria) have their own DNA – mtDNA – apart from the main DNA in the cell nucleus, and is inherited solely through the egg of the mother. Thus, small changes in its sequence of base pairs, accumulating over time, may be used as a kind of molecular clock. Mothers who leave no daughters disappear from the mtDNA record, but others continue it, so it possible to construct a kind of family tree of all humanity.

This theory was immediately and hotly contested by paleoanthropologists, such as Thorne and Wolpof.[16] They and others raised numerous serious objections to the OAF hypothesis, first, that the mtDNA method was seriously flawed (which was true), and second, that fossil evidence indicates that the various peoples around the world arose from prior populations in the same regions, going back as much as a million years. This conflicts with the OAF hypothesis, since the OAF claims that all prior populations were entirely displaced by the wave of the modern species, a mere 60,000 years ago (60 ky), give or take. That is to say, the fossil evidence suggests that modern Asians, for example, arose from Asian *H. erectus* because the old *H. erectus* in that area have many features which persist in the modern population. Similar arguments apply to other regions.

Now, fast-forward to ca. 2007, by which time the OAF hypothesis had gained more strength, and its opponents, the "multi-regionalists, seem to be fighting a losing battle. Among the more persuasive recent papers is by Manica et al, who sequenced entire mtDNA genomes (instead of just small parts of it) of populations around the world, and at the same time performed cladistic (or "morphometric") analysis of skull features of those people. That work demonstrate a clear gradation of both measurements leading out of Africa and spreading worldwide.[17] The highest degree of variation in both mtDNA and morphometric skull features ("races") occurs within Africa, and the least variation occurs at the terminal points, such as among Asians or Australians. So, it's beginning to look like the OAF hypothesis is here to stay.[18] Indeed, dozens of high-profile articles have recently appeared, and continue to appear, which further support the case. It is hardly necessary to cite them, since this is now common knowledge.

The dust has not yet entirely settled, however. There is still a lot of debate,[19] especially concerning details. For example, modern humans escaped Africa at least twice, but it seems that the first escape had little consequence, although it may explain the recent find of an early *H. sapien* in China, with signs of interbreeding with H. erectus.[20] (The reason why escape was difficult is that Neanderthals blocked most routes out of Africa, discussed in a later chapter.)

However, most of the debate now focuses on matters such as the particular route taken, and the sequence of places settled by the modern humans. It looks like the route taken out of Africa was through the Levant and then mainly coastal from India to China.[21-23] This is intriguing since those were the regions where the classical civilizations later appeared, beginning around 8 kya, when large-scale agriculture arose, capable of supporting high population densities. The "primitive societies" were those which wandered away to comparative isolation, remaining more-or-less less frozen in time, e.g., Australia, the Arctic. The Americas, or "New World," was populated relatively late, about 17 kya. Nevertheless, this was well before the rise of any major civilization, and hence, before the kingships, writing, pottery, or other marks of civilization. No prior *H. erectus* has ever been found in the New World.

3. EXPLAINING THE BOTTLENECK. It goes without saying that the origin of marrying (and kinship, exogamy) was a major event in the history of our species. It might also go without saying that we are suggesting that it was the fuel that propelled us out of Africa. In this article, we demonstrate how the marrying hypothesis can explain yet another major conundrum concerning the OAF hypothesis.

It so happens that among the findings of the genetic studies undertaken in defense of the OAF hypothesis was the discovery that the initial population giving rise to all of today's peoples must have been extremely small. It is beyond the scope of this chapter to attempt to explain exactly how this conclusion was reached, but rest assured that it is based on reasonably secure genetic science.[24] The big surprise was that the original population turned out to be about 1,000 people, or possibly less.[25] This result, found and confirmed by multiple investigators, sparked a flurry of efforts to explain it. In view of this finding, it was widely concluded that our species was on the brink of extinction!

How to explain that? Well, one theory that was popular for some years was that a big volcano, Mount Toba, which exploded 74 kya, was the culprit, causing disastrous climate change and massive die-off.[26] That theory failed to hold up, for several rather obvious reasons, but other theories soon replaced it. Thus, one expert has described how the few survivors of our species took refuge from a suddenly harsh climate, huddled by the sea on the coast of South Africa[27].

However, our marrying hypothesis offers a very simple and entirely different explanation. There was no massive die-off. Instead, what happened was that the original two-clan society ceased interbreeding with surrounding populations and expanded, fissioning into multiple bands, all descended from its original founders of perhaps a few hundred. This was probably linked closely in time with the above-mentioned advance in language. Within a few dozen generations, or a few thousand years at most, this population would have all of the earmarks of having survived a massive die-off. But it wasn't a die-off. There was no big disaster. It was simply an elite population that married only among themselves, according to the clan rules.

It was one of those groups that made the final and successful foray out of Africa, founding all of today's living humanity, spreading matrimony with it. Eventually, we must surmise, all of Africa was also replaced with marrying societies (since they all get married), but they did not necessarily all descend from the original innovators, they may have copied the system, or were adopted into it as additional clans by one of the marrying bands, or other scenarios, as described in the next chapter.

4. THE FINAL PUSH. It is known that modern humans had ventured out of Africa once before, but were apparently blocked by the Neanderthals. The final and successful push occurred some 65 kya. How did they succeed, while their predecessors failed?

Although the Neanderthals are the topic of a later chapter, it is necessary to explain that they were a branch of *H. sapien* who originated in Africa around the time of the first *H. sapiens*, 300 kya, and apparently acquired their distinctive features within Europe and its surrounds, surviving through some very tough conditions there. Most authorities consider them to be a subspecies, *H. sapien neanderthalesnsis*, as distinct from the "anatomically modern humans" (AMH), *H. sapien sapiens*, who penetrated into Neanderthal territory early in their exodus, *ca.* 50 kya.

The immediate question is how were the modern humans able to penetrate the Neanderthal stronghold, and to eventually displace the Neanderthals? It was certainly not by brute strength, for the Neanderthals were much bigger and tougher (but the modern humans could outrun them[28]).

Here again, the marrying hypothesis can help explain this success. For the marrying people by now consisted of multiple bands of cooperating clans, united by strong kinship ties. It is not known if the Neanderthals had strong inter-band alliances, but as a guess, each band probably stood alone, proud and independent. Of course, there are many other theories, such as the possibility that the modern invaders possessed the bow-and-arrow, but that is controversial.

Most expert commentators tend to downplay, or not mention at all, the occurrence of outright violence and warfare between Neanderthals and the newcomers, but sporadic actions of that kind is certainly a distinct possibility. There is plenty of evidence of violent death, although it is rarely clear who the killers were. The scenarios usually given for the ultimate extinction of the Neanderthals refer to "competition for resources" but it is difficult to imagine such competition without violent confrontations such as concerning territorial boundaries. Hunting peoples tend to jealously guard their territories, on pain of death; hence the newcomers must have possessed a powerful advantage in order to bring about the demise of the Neanderthals.

The clan-type of social structure, with its inherent permanent alliances between bands of related clans, must have been the decisive element in such confrontations. Then too, there may have been a less tangible but also decisive element, known to military leaders as "morale." Indeed, the more recent collapse of primitive societies world-wide when faced by the Europeans was not only a matter of guns and disease, it was also the destruction of their gods, their faiths, their cherished beliefs, when faced with the greater knowledge of the world possessed by the European conquerors. More will be said about such matters in the chapter on Neanderthals.

5. CONCLUSIONS. Although more remains to be said, we have shown in this chapter how the marrying hypothesis can tie up many loose-ends in the OAF hypothesis, including the mystery of the population bottleneck at the dawn of the modern species. We also suggest, but more tentatively, that the failure of earlier efforts of modern humans to expand out of Africa might signify that some additional pivotal advance had not yet taken place. It is possible that the key advance creating our species, 300 kya, was true human language, and that the marrying society arose only much later, ca. 65 kya, being the final spark that ignited the successful thrust out of Africa.

7.

The System Speads

1. PREAMBLE. It is obvious that the marrying system *did* spread, since all humanity does it. This book differs from the conventional accounts, however, in proposing that the origin of marrying was a singular historical event, very near to the root of the spread of all living humanity worldwide. Whereas the usual accounts of the African exodus generally ignore the whole question of matrimony, leaving it at some unspecified point in dim antiquity, we hold that it was an absolutely seminal event, the very well-spring of human culture, and was the spring-board behind the ultimate exodus of modern mankind out of Africa. In this chapter, we conjecture on some of the intellectual and interpersonal dynamics that drove the spread of the marriage-exogamy-kinship (MEK) custom.

2. THE HOW OF IT. It goes without saying that the marrying society – the MEK complex– must have enjoyed certain survival advantages, since it somehow eclipsed all other contemporaneous peoples. The usual term would be "evolutionary fitness." But exactly how did this happen? As the reader may divine from earlier chapters, we prefer our explanations in terms of concrete scenarios, as opposed to vague and generic arm-waving terms like "competition for resources" and the like.

That is to say, those of the materialist persuasion can doubtless dream up all sorts of Kiplingesque benefits of matrimony that enabled it to out-compete neighboring groups – and they have, as noted earlier. But this writer is not of that persuasion, holding instead that *ideological* and *intellectual* concerns are of equal or even much greater importance in molding, fixing, and defending cultural innovations. This should be self-evident from the fact that so many people have been willing to suffer and die, even by unspeakable torture, in defense of their ideals, their principles, their religions, their honor – their *beliefs*, in a word. Thus, we side firmly with Levi-Strauss and the French tradition of anthropology on this matter.

But exactly what kind of ideological concerns underlie the success of the marrying tradition? Well, for openers, this institution stood for *fairness* and *justice* – the guarantee of a personal spouse for everybody – and for *human equality*. These are earmarks of a political ideology. Although we do not know what the pre-matrimonial mating system was, it definitely did not systematically assign a wife to every man. It may be noted, by the way, that the matrimonial system flies in the face of orthodox evolutionary theory, for there is no mechanism in the matrimonial system to ensure that "good genes" are selected, since *everybody* has the right to a spouse and to bear children, not just the "fittest."

That said, let us now climb back into our trusty time-machine and revisit that pristine river valley where we first observed the origin of marrying, and check up on our innovative friends a few centuries later. The honeymoon is clearly long over, yet the system persists. A first sign of their success is that they have fissioned into a few dozen additional bands, radiating out from the original source in all directions. Each is almost a carbon-copy of the original, although some deviations are already conspicuous, such as in the names of some of their clans, i.e., totems. Some or most bands near the original source retain their initial clan identities, inter-marrying with the original group and its other splinters in accord with the same system of rules.

Of particular interest to us, however, is the impact of the marrying people on the surrounding non-marrying bands. Now, it is certainly possible that the marrying peoples obliterated some or many of their neighbors by violent means, made possible by the powerful alliances of the clans. That is to say, although individual band size cannot increase for hunter-gatherers at a specified level of resource exploitation, the clan size can increase, almost without bound, as the founding band fissions into many. On the other hand, although the clan system was the basis for *stable* alliances, it must be expected that non-marrying people also had defensive alliances, albeit more ephemeral, and so would not have been push-overs when push came to shove.

But conquest by violent means is not a necessary assumption for the spread of the system. One might envision also *assimilation*. How would that work? It would work mainly by its intrinsic appeal to the majority of those exposed to an MEK society newly appearing in their region. In all likelihood, the pre-matrimonial bands were dominated by a few alpha-male types, not as tyrants, of course (for reasons earlier explained) but as benevolent *de facto* kings, perhaps not so different from silverback gorillas. Nevertheless, as also earlier pointed out, the pre-matrimonial arrangement would almost certainly have entailed a good deal of sexual frustration, resentment, and discontent for a sizeable fraction of the group. Therefore, following contact with the revolutionary marrying people, and learning how their system worked, more than a few pre-matrimonial bands might well have duplicated the system for themselves, with or without the approval of their leaders. An equally viable alternative is that the marrying people might have actively recruited or invited neighboring bands into their systems, as additional clans.

Here it must be mentioned that adding a third clan to accommodate an additional band entails a considerable degree of intellectual gymnastics. It is beyond the scope of this outline – and the writer's expertise – to discuss the intricacies of marrying within multi-clan systems. And yet, multi-clan systems have been the rule, not the exception, in historically known societies, particularly in Australia, and the rules of marrying and descent (clan identity of offspring) are exceedingly complicated and variable.

It seems a fair surmise that multi-clan systems could also have arisen by the fusion of two or more marrying societies. They could also have arisen by the fragmentation of one or more clans *within* a society, since it is well known that serious frictions often arise within a clan, so that breaking one into two clans (or sub-clans) might often have been a wise expedient in such circumstances.

3. MISSIONARY ZEAL? Let us go further with the ideology theme, suggesting that the marrying people might have vigorously promulgated their system to pre-existing bands in the neighborhood, or encountered along the way of their expansion outside of Africa. For after all, they had discovered a brilliant new political system, as it were, no less than the world's first socio-political ideology! Indeed, to this very day, we still subscribe to it. So, little wonder that they may have harbored a certain pride, the same as that exhibited by all empires and religions, bordering on missionary zeal, fueled in equal measures by lust and intellectual conquest.

This has testable implications, however, which appear to be at variance with the facts. In particular, if we assume that the spreading marrying people assimilated pre-existing peoples in their global march, most of whom would have been of the *H. erectus* spectrum, traces of this should remain in anatomical and genetic features. But the DNA evidence thus far shows only traces of input to the gene pool, although some major questions remain, to be discussed presently.

4. THE LANGUAGE QUESTION. Although beyond the scope of this book, the author has made what he believes to be a major discovery about the nature and structure of human language, quite different from any existing conception, and which implies that true human language (TLH), like the marrying system, was practically an overnight invention. Granting that possibility, and the fact that language was an essential pre-requisite for the origin of matrimony, then it seems likely that the marrying system emerged shortly after the advent of THL, perhaps within a few centuries, but possibly only after a dozen or more millennia, and in the same region.

If such was the case, then we have a very different set of circumstances, for this original society was unique not only in its marrying system but also in possessing true human language. If so, they may well have looked upon their non-speaking, non-marrying neighbors with the same contempt with which our civilization has looked upon animals, and so may have had no qualms about exterminating them, and no temptation to assimilate or "convert" them. This scenario seems to be most consistent with the DNA evidence, which generally shows little input to the original gene pool from the out-of-Africa people, with some caveats discussed next.

5. ASSIMILATION? It was earlier mentioned that the opponents of the out-of-Africa (OAF) hypothesis, being the multi-regionalists, had based much of their arguments on morphological similarities between existing peoples at various places and archaic fossils found in the same region. We earlier mentioned the oriental features of the Peking Man, but another poster-child of the multi-regionalists has been hints of *H. erectus* features in living Australian aborigines – Abo's, for short. Also earlier mentioned was apparently conclusive evidence of the absence of any such assimilation.

One of the very few benefits of getting old is a better perspective on "the latest scientific advances", by which I mean having witnessed a succession of headlines about genetics, many if not most of which turned out to be premature, or over-simplified, or flatly wrong. Coming to the point, it now turns out that our DNA is not quite so pure as claimed as recently as a few years ago, but is actually sprinkled with varying degrees of admixtures with more archaic hominids, depending on geographical region. Thus, we now see headlines like, "Out of Africa, into Bed with the Locals," [1] and "Our Hybrid Origins." [2] These recent findings were spearheaded by work by Svante Paabo on the Neanderthal genome, finding input from several hominid branches.[3] Indeed, a completely unexpected and previously unrecognized early *H. sapien*, dubbed *Denisoven*, has come to light, found in a cave of that name in Siberia, distinct from Neanderthals, dated to about 35 kya, and showing up in some modern human DNA.

Coming back to the Australian Abo's, recent literature shows this to be of continuing interest. A limited genetic analysis (Y chromosome and mtDNA) of Abo's showed no evidence of archaic input,[4] but it now appears that only whole-genome analysis would suffice to decide the issue. A comparative study of Abo crania with older *H. erectus* from Java failed to show common features,[5] but limited interbreeding would not be expected to exert gross change in morphology. Another analysis of Abo crania confirmed their distinctive uniqueness,[6] and a more recent (2011) overview of the subject concludes that the robustness of the Abo cranium remains to be explained.[7] A series of Abo skulls dating to about 20 kya showed two distinct types, robust and gracile, another puzzle.[5]

We conclude that the last word has not been said. It may also be mentioned that modern genetics is still in its infancy. One authority commented that "centuries" will be needed to fully understand how the genome works. At least two huge and revolutionary discoveries about genetics – the role of micro-RNA, and epigenetic marks – came to light in just the past decade, and rest assured that more are on the way. Highlighting our ignorance is the fact that we still have no idea of how the genes control such features as facial and skeletal morphology. Clearly, the last word has not been said on the issue of limited assimilation, although it must be admitted that the mitochondrial DNA evidence is hard to refute.

6. VARIATIONS ON A THEME. The next problem is to understand the rise of the diversification of marrying practices and related social features. The Australian scene is most readily understood in terms of our MEK hypothesis because, despite the complexity of its numerous marrying and kinship systems, essentially all of them are plainly clan-based, and therefore can easily be imagined to have arisen by mechanisms such as those posited above, or other events that complicated the systems during the intervening 50 thousand years. These "other events," aside from clan fragmentation or group fusion, might have included deliberate intellectual tampering to make the system more consistent with, for example, religious beliefs, or a mystical sense of geometrical symmetries.

Such factors were evidently important to Australians, who much enjoyed expounding on the fine points of their social systems. As earlier mentioned, Australia is particularly supportive of our thesis, since it may be thought of as a sort of 50,000 year time-capsule, consistent with genetic evidence showing little or no influence from outside of it.[4] The central highlands of New Guinea is another good case in point, since it too, was entirely clan-based, and insulated from the flux of the coastal populations by high mountains. Known archeological evidence of habitation there dates back 49,000 years.[8] By the way, this fascinating region is the earliest known example of agriculture, apparently having been invented independently there. Hence, it can be no accident that these places – Australia, New Guinea's highlands, most of the Melanesian Islands – are all entirely clan-based, in contrast to so many other places, where great diversity of MEK systems exist (but almost always interspersed with clan-type systems).

At the other extreme is the complete disappearance of clan-type structure, especially in large civilizations such as our own. The major civilizations, which were all agrarian, arose by assimilating (or exterminating) neighboring peoples, in which circumstance it becomes practically impossible to keep track of clan identity, beyond that of the extended family. That is to say, in our society the "family name" is the sole surviving trace of the original clan – that, and the rule that you must marry outside of it.

Between these extremes is the vast diversity of marrying systems seen elsewhere in the world, such as in pre-Columbian North America, i.e., among the oxymoronic "American Indians," better called native Americans, or *Amerindians*. The Americas were populated from Asia some 30,000 years later than Australia, so it is safe to assume that its founders were privy to a great deal of additional cultural lore accumulated from the Eurasian mainland, including a greater sense of their potential for change, variation, novelty. The Hawaiian Islands were populated much later yet, ca. 7,000 years ago, and brought with them the comparatively recent kingship type of social order. To explore the question of variety, we need to get back in our time machine.

7. TREKKING TO NOVELTY. We set the controls at 16,000 years ago, and hover over Montana's lovely Bitterroot Valley. As luck would have it, we find people there at the then-current southern limit of their expansion. All suitable lands to the north are already inhabited to the maximum carrying capacity. What we first observe is the usual cloning of the current society into multiple outposts, fanning out as more-or-less independent bands, but united by clan identities, all of which periodically congregate for weddings, funerals, feasts, and general confabulation, as observed in, for example, Abo's known as the Tiwi, living on a large island off Australia.[9] The original clan system is fully intact, albeit complicated by some additional clans and perhaps sub-clans, or other divisions.

Then we chance to witness an important event, namely, the departure of a group of *malcontents*. It is not immediately clear if they were forcibly exiled, possibly even fleeing for their lives, or departed of their own volition, but one thing is clear, they are not happy with the state of things back home, and so they have set off south, never to return.

We learn that the disconsolation of the leaders of this faction actually stemmed from rivalry about status, leading to fights and their expulsion as trouble-makers; but they don't see it like that, blaming their problems on other matters, *viz.*, the *unfairness* of the system back home, and the stupidity of certain marrying rules, as well as questions about which god or goddess is most deserving of their honor. Day after day they walk, excitedly discussing as they go the "ideal society" that they intend to found, as soon as they find the right spot. Endless details are subject to change. For example, since clan membership must be either patrilineal or matrilineal, whichever it was might be seen as unjust, or backwards, or stupid, and so could be reversed. Likewise, the clan names (totems) could change, as could the clan to which the children are sent to live (matrilocal or patrilocal), and customs for the inheritance of certain privileges, and procedures for divorce, etc., any or all of which might be blamed for their troubles, and so must be changed.

And so it was that after more than three weeks and nearly 200 miles of exuberant walking and talking, they came upon a sweet and blessed place on a fast-running stream, near a lake. The stars and planets are favorably aligned, and dancing birds are further omens. This will be their new base camp, their Promised Land, the heart of their new territory, teeming with game and other good food. They drop to their knees for a moment of prayerful reverie, then begin immediately to implement their New Society, radically different in being matrilineal rather than patrilineal, and different in several other respects, too. However, as anthropologists well know, human societies are extremely holistic, in the sense that all beliefs, practices, folklore, customs, habits, diet, etc., are inter-twined by logical networks, i.e., are self-consistent. Indeed, they may be likened to *logical* systems, replete with axioms, corollaries, and innumerable theorems, so that when you alter any one of the foundations, the end result is an entirely transmogrified system.

So it was with our intrepid band of founders of a New Society. Within a single generation, numerous ancillary modifications were found to be necessary, few of which were originally anticipated. The same phenomenon occurs when a large corporation decides to re-organize: the intended changes bring with them a seemingly endless sequence of unanticipated consequences, and new problems to solve.

The bottom line is that within a few generations, this new society might be practically unrecognizable as having sprung from its actual parent society. These principles, too, are patterned on Levi-Strauss, and the writer has applied them also to languages and to his psychogenic theory of biological evolution. It is important to point out that true novelty is extremely rare in the history of social systems and technologies. What is not rare is variations on a theme. That is to say, the New Society witnessed above, although imagined by its founders to be exceedingly revolutionary, was in reality accomplished only by inverting, reversing, opposing, or contrasting certain major features of the parent society, resulting in a subsequent cascade of additional modifications, until at last, after perhaps a dozen generations, the system was perfectly coherent, logically self-consistent in its every detail. The process has been likened to turning the window of a kaleidoscope: the same crystals, tumbling into new concatenations, exhibit an entirely different appearance.

To all outward appearances, it seems utterly transmogrified. When all the loose ends have been tied up, and the myths and gods brought into alignment with new system, then the system is finally a logically coherent whole. It appears perfect. When that end-point has been reached, its members are trapped within a logical sphere that appears so perfect that no reasonable alternative to it can be imagined, there is no room for further improvement. They have reached the end of their developmental trajectory. Only the malcontents, the misfits, the disenfranchised, the exiles, the *outsiders* are in a position to see it from the outside, and so to foment revolution.

Indeed, it appears that on numerous occasions, founders of new societies decided to abolish the clan-type system altogether, often for the simple reason that clans are commonly sources of corrosive social discord, rivalries, friction, dissent. Of course, abolishing clans leaves a kind of vacuum calling for some sort of alternative system for maintaining kinship structure and the basis for the exogamy rule.

It may be added that our little band trekking through the Bitterroot Valley enjoyed a rare and extraordinary opportunity to actually set off into the wilderness and set up a new society however they wanted it. For the then-virgin continent was soon filled to capacity. But it so happens that there are other mechanisms for changing the rules, for generating societal variegations.

8. NEIGHBOR CONTRASTS. One of these other mechanisms shall be termed *neighbor contrast*. The author was first struck by this theme by surveying the cultural anthropology of New Guinea, later by the study of language differences between neighboring peoples, and finally by Levi-Strauss' 5-volume analysis of local myths among neighboring groups. Very briefly, it appears that every distinct society *actively seeks to differentiate itself* from its neighbors by the systematic exaggeration of otherwise trivial differences in its customs, including matters of diet, language, religion, marrying customs, *ad infinitum*, as compared with its neighbors. This, too, could lead to profound alterations of marrying and kinship rules.

Thus, there are plenty of mechanisms to account for social changes, including the widespread disappearance of clan-type structure. Back in the 1960s, when this writer was a serious student of anthropology, the big question was accounting for such changes; and the big answer, back then, was *economics and environmental forces* – that is, a Darwin-inspired philosophy of materialism, or "social ecology." [10] Under that oppressive paradigm, everything was caused by external forces; the people have no say in what they do.

Those efforts had some interesting successes, and nobody can doubt the importance of economic factors to human survival, and in altering the complexion of daily life. However, that program of social anthropology was largely a failure, in the sense that it was not generally possible to persuasively account for the variety of social systems by those factors alone. (Let us add that much the same is true of the general theory of *biological* evolution: it is not possible to explain the variety of species by material considerations alone.) We must therefore conclude that psychological or *intellectual* machinations are the true Rosetta Stone for comprehending the varieties of social orders (and animal species).

The bottom line of this chapter is that the observed variety of marrying and kinship systems around the world does not in the slightest challenge our central hypothesis, which is that they all emanated from an original two-clan system. A further item of powerful support was earlier mentioned, namely, the very high prevalence of the cross-cousin marrying preference around the world, in societies with *no surviving hint of clan-type structure*. This preference has presumably persisted from the ancestral clan-type structure to preserve some semblance of order or subliminal rationality in the process of mate selection following the dissolution of the clear directives inherent in clan-type societies.

8.

Did Neanderthals Get Married?

1. MEET THE NEANDERTHALS. Of all the hominids, none is more popular than the Neanderthals. The name comes from the Neander valley in Germany, where the type specimen was found in 1856, in a quarry, the first prehistoric hominid ever found, although two others had been unearthed some years earlier (one in Belgium, one in Gibraltar). Darwin's theory of evolution was not yet known, so there was much early confusion about what to make of the bones. Indeed, the workers who discovered them mistook them for the remains of a bear. Around 1900 Boule published a careful study of a specimen found in France which study was long respected as the definitive account, giving rise to the persisting stereotype of a shambling, hairy, gorilla-like ancestor. Only many decades later was it found that Boule's account was grossly distorted. Among other things, it turned out that the specimen studied by Boule was an old man, deformed and crippled by arthritis.

It is now known that the Neanderthals were true *Homo sapiens*, usually designated as a subspecies, *H. sapien neanderthalensis*, as distinct from ourselves, *H. sapien sapiens*. Their remains are often associated with a distinctive style of stone tools, dubbed the Mousterian industry. Genetic analysis indicates that they split away from early African *H. sapiens* about 250 kya, and subsequently occupied Europe and surrounds from about 220 kya until their extinction about 28 kya. Their distinctive features must have arisen in that interval.

Back in the 1960's, many questions about the Neanderthals were hotly debated, and many of the same questions are still debated today, still hotly. A good perspective can be had by comparing articles on the Neanderthals in the *Scientific American* from 1957,[1] 1979,[2] and 2009.[3] On the other hand, a great deal of progress has been made, not least of which has been the discovery of numerous additional specimens and their associated artifacts, all now dated with increasingly reliable accuracy.

Their brains occupied a volume of about 1600 cc, which is somewhat larger than our own (mean 1400 cc), but the Neanderthal head was differently shaped, more like an egg lying on its side, whereas ours is like an egg stood on end. This gave them a low and sloping forehead and a kind of bun at the back of the head. Lieberman and Crelin advanced an attractive hypothesis to account for this shape, namely, that our higher dome was pushed up to make room in the mouth for improved speech capability.[4] If true, that would account for a lot, but unfortunately, they later retracted that hypothesis as inconsistent with the facts.[5] Of course, Neanderthals have numerous distinctive features, from head to toe, such as prominent brow ridges, distinctive teeth and jaws, and endless details appreciated only by specialists.

The Neanderthals were definitely a rough and tough bunch. Their bones were much sturdier and heavier than ours, with muscles to match. They were hunters of really big game, not only of deer, elk, horses and reindeer, but also of bison, mammoths, rhinos, even giant cave bears, all of which are formidable foes. Today's hunters, with their high-powered rifles, are wimps by comparison, hardly deserving the title of "sportsmen." It was apparently a hands-on sport, for a great many Neanderthal specimens show evidence of healed fractures resembling those of rodeo riders. One suspects that they enjoyed the challenge. Their many healed injuries attest to good nursing care when laid up. Many elderly and infirm specimens also indicate compassionate care.

The Neanderthals must have been highly resourceful to occupy Europe during periods of heavy glaciation, which obviously required warm clothing, shelter, and provisioning food for winter, challenges not faced in

Africa. Their remains and artifacts are best preserved in caves but they must have had other types of shelter as well, perhaps tents.

2. THE ARTSY INVADERS. The "anatomically modern humans" (AMH), which we shall call simply *Cro-Magnons* (after that type specimen, *ca.* 30 kya), entered Europe for keeps about 40 kya, and somehow coexisted with Neanderthals for the next 10,000 years. The last known Neanderthal was found in Gibraltar, dated to 28 kya,[6] but some other pockets appear to have survived a bit longer, in inhospitable Siberia.[7,8] As earlier mentioned it is known that modern humans had attempted earlier incursions but were apparently blocked or driven back by the Neanderthals. How did they succeed this time?

A key distinguishing feature of the Cro-Magnon's artifacts is the appearance of art and ornamentation, all said to be evidence of "symbolic thinking," whatever that means. The best known evidence consists of the cave paintings, mostly dated from about 33 to 15 kya, all of which are associated with the fully modern Cro-Magnon type. Outstanding reproductions have been published,[9] some of which are timeless masterpieces by any standard, and were said to have become more refined and sophisticated over their 15,000 year span.[10] However, a recently found (1994) example was dated to 31,000 years and is so magnificent that some have doubted such great age.[11] Some are found at great depths and could only have been painted with the aid of oil lamps, hinting at mystical motives. The walls often include patterns of squiggles, dots, hand-prints, crescents, wavy lines and zigzags, long believed to have been mere graffiti, but the idea was recently floated that they might actually be a written proto-language.[12] The patterns had been classified already in 1968[10] but that author did not suggest any sort of writing system.

Second on the list of artistic artifacts are the carvings and statuettes, best known of which are the small and rotund "sex goddesses," with exaggerated sexual parts. Dozens have been unearthed, and the most recently found may be the oldest yet.[13] The ubiquity of these figurines adds to other evidence of connected cultural features over a wide area. Marks found on a carved antler may have been a kind of lunar calendar. A flute fashioned from bird bones dated to 32 kya is said to be the oldest undisputed evidence of music.[14] Necklace beads were often found. Improved styles of stone tool making, of greater diversity for special purposes, is also associated with the

Cro-Magnons. However, as mentioned earlier, research in the last decade is making it increasingly clear that by no means all of this evidence of cultural advance originated in Europe, since related finds in Africa going back 100 kya or more have been unearthed, e.g., very old shell beads in Israel.[15]

On the other hand, it is important to emphasize plenty of evidence that Neanderthals were not devoid of higher sensibilities. Ralph Solecki studied a Neanderthal burial site at the Shaidar cave, finding the deceased to have been lovingly interred with artifacts, painted with red ochre, the grave strewn with flowers, as inferred by pollen analysis.[16] Terry Hopkins and colleagues are among those who argue persuasively that the level of cultural attainment achieved by Neanderthals is chronically under-valued.[17] Zilhao and colleagues make the same point in the title of their report, "Symbolic use of marine shells and minerals ..." by Neanderthals.[18] These observations bring us to the next article.

3. THE DIM-BULB HYPOTHEIS. To explain the apparently more sophisticated culture of the Cro-Magnons *vs*. Neanderthals, the opinion has been ventured that the latter were not quite as bright as some people imagine themselves to be. This is in contrast to the opinion that purely cultural factors, such as those possessed by the marrying people, could equally well explain the disparities. This boils down to the old nature-vs.-nurture debate.

By way of preface, it is the author's impression that those who turn to genetics or brain-wiring to explain the success of one group compared to another are themselves mentally defective. This is because everybody knows, or should know, that the success of great civilizations – Egypt, Greece, Rome, Islam, China – is not the result of genetic superiority, being always ephemeral, each lasting a few centuries, only to be replaced by another. Genetics has nothing to do with it. Furthermore, even within a high-achieving civilization, such as our own, nearly all of the pivotal advances were made by a tiny few individuals. The average American today has not the faintest clue about how an ordinary radio works, much less Einstein's relativity theory. Anthropologists have fought long and hard against racism by demonstrating the enormous power of culture, as opposed to innate mental capabilities.

To illustrate, let us briefly consider the hypothesis of Wynn and Coolidge, which seems to be gaining proponents, judging by the fact that it was recently featured in the prestige journal, *Science*.[19] The following quotation sets the stage:

> "Paleoanthropologists trying to answer this question [of the extinction of Neanderthals] are slowly moving toward a consensus, although it is still not unanimous. They believe that these later *Homo sapiens* must have had certain cognitive abilities not shared by Neanderthals." [20]

Euphemisms aside, they were dim-bulbs. Wynn-Coolidge suggest that we have an enhance capacity to use symbols. They claim evidence of improved ability of the Cro-Magnons to plan and strategize, compared to Neanderthals. They discuss at length the many advances made by the new invaders, along the lines we summarized above. Then they get more specific.

> "In particular, there is no indication of continuing innovation: Neanderthals made the same kinds of tools for 200,000 years. ... They practiced little, if any, conscious [sic] invention ... suggest that they did not have a second kind of modern problem-solving ability, one known to psychologists as 'executive functions' ... *working memory*. ... Neanderthals never acquired that mental equipment, only we did." [20]

First of all, it is not correct that they made the same kind of tools for 200 ky.[17] There follows a lengthy account of how this kind of alleged higher intelligence is supposed to work, a description of some brain anatomy, brain imaging studies, and so forth, much of which is widely dismissed by critics as pretentious, bordering on junk science. As for signs of progress or innovation, nobody back then showed much innovation, nor have most societies known in historical times. Many have existed without significant change for tens of thousands of years. Why? Because they were dim-witted? Lacked some key gene mutation? Were biologically inferior? Of course not. Many Australian Abo's are now college professors. Hence, their position amounts to racism, by using the same flawed logic that the White Man used to justify the genocide and slavery of primitive peoples all over the world: high-falutin rationalizing, sans especially genetic inferiority. We now know that primitive societies were in no way intellectually defective, yet they failed to invent things like steam-engines and battleships. Likewise, we submit, the failure of Neanderthals to pain pretty pictures as the girly-men did does not in itself make them retards.

Accordingly, our bottom-line conclusion is just this. Yes, the Cro-Magnon invaders of Europe exhibit a good deal or cultural novelty, but the more parsimonious explanation is some kind of *cultural* invigoration, not innate giftedness. What kind of cultural invigoration? Well, the invention of matrimony and kinship-based clans is a good candidate. The invention of true human language is another. Continuing now with the Wynn-Coolidge hypothesis, they address the question of how we managed to get so much smarter than Neanderthals. This being the 21st century, their answer is predictable: a genetic mutation of course!

"We suggest that a genetic mutation in the ancestors of modern humans caused a re-wiring of brain neurons that resulted in [more smarts]. ... For example ... gene known as FOXP2 ... [or] micro-cephalon (MCPH1) ..." [20]

Sprinkled through their text are allusions to Haldane's population genetics, the new field of "cognitive archeology," and so on. Now, the FOXP2 gene is indeed likely to have something to do with our language capacity, and differs from that in apes. But as Wynn-Coolidge acknowledge, perhaps ruefully, the fully modern form of this gene has been identified also in Neanderthals. As for MCPH1, it doesn't much matter if that gene – and its promoters, regulators, etc. – is exactly the same in us and Neanderthals, because nobody knows quite what it does, and besides, it exists in many variant forms even among us.[21] Likewise for the rest of brain genetics.[22,23] We don't have a clue about what genes, if any, control features of the skull and face, foot bones, inner ear, etc., much less the detailed wiring of the brain. The Cave Men knew as much about such things as we do, and possibly more, judging by the fact that they made us what we are.

That brings up another general point deserving a few lines, namely, the prevailing dogma that genetic mutations lie behind every detail of behavioral or cognitive change. This is exactly backwards. What actually happens is that people make choices or discoveries, and then the genes arise to serve the new directions. Take the drinking of milk. Most of the world's peoples, such as Asians, cannot tolerate milk after weaning, because the gene for the enzyme (lactase) needed to digest milk sugar (lactose) becomes inactive. However, a goodly number of people rely on milk as a staple throughout their adult years, and to that end, have acquired genes which maintain the enzyme's activity. How did this happen? Did they wait a million years to get that gene by random mutations, and then suddenly acquire an unquenchable thirst for milk? What actually happened was that they started drinking milk and – Presto! – the needed genes popped up. Not only that, it happened in at least four different human groups (Northern Europeans and three African tribes), each of which produced a different genetic mutation to do the same trick.[24] Milk presumably offered health benefits which endowed its drinkers with more and stronger babies, resulting in spread of the mutation through the local population. But even that is open to serious doubt. They may have taken up drinking milk for purely cultural reasons, perhaps for mystical or magical or religious reasons, so that natural selection had nothing to do with it.

A different example is the white skin of people living in northern latitudes, including Neanderthals. Here again, the Neanderthals got their light skin color (and orange hair) by different genetic mutations than those we have for light skin.[25-27] The practical purpose of light skin in the cold north, where a lot of clothing must be worn, is to maximize absorption of sunlight to produce vitamin D3.[28] In this case, the white skin may well have been the product of natural selection. However, the spread of this mutation may well have been greatly hastened by cultural factors, meaning that light skin and hair may have acquired value as beauty, promoting sexual selection for these traits.

In any event, we conclude that the hypothesis of Wynn and Coolidge – that the alleged higher intelligence, or "working memory," of the Cro-Magnons resulted from a fortuitous mutation that rewired their brains – is not only pure speculation without rational support, but is exactly backwards. The more plausible scenario is that the early *Homo sapiens* were doing things and thinking things, and trying to talk about things, which required a wiring upgrade. The plain truth is that, as far as we know, the Neanderthals were our mental equivalents in every way. All the rest can be explained by culture.

4. THE CROSS-BREED QUESTION. Coming now to the title question of this chapter – did Neanderthals get married? – the short answer is *no*, and this for the reason that the origin of matrimony was precisely the event that launched the exodus from Africa to the four corners of the world. Only that hypothesis explains all the facts – the worldwide prevalence of marrying, the population bottle-neck, success at invading Europe, and a host of other questions. Nevertheless, speaking as an inveterate Neanderthal wonk, the question of interbreeding is quite interesting, and bears on our thesis.

For more than 50 years, one of the hottest debates was whether the Neanderthals inter-bred – or married? – with the fully modern Cro-Magnon newcomers. Of course, if they really were a distinct species, as some assert, then by the classical definition of a "species" no viable offspring could even-tuate. Beginning in the 1990's, it became possible to isolate snippets of DNA from various Neanderthal remains, such as teeth or bone marrow, and by 2006, the "Dawn of Stone-Age Genomics" was announced.[29] There were a number of false starts and premature headlines, caused by things like sample contamination and inadequate methods. In 2008, a review of all work to date concluded confidently that "there is no evidence of any Neanderthal genetic contribution to any humans now living." [30] Case closed? Not quite.

Fast-forward to 2010, when it was established that segments of Neanderthal DNA are indeed present in modern humans, at least in some European and other groups, but not in Africans.[31] Not only that, but another population, dubbed Denisovan, of central Asia, which is appar-ently a Neanderthal-like subspecies, also entered the human germline, and is found at least in modern Melanesians. The picture is quite complicated[32] but one thing is clear, there definitely was some early interbreeding, albeit very limited.

Of course, we can never know about the many possible interactions or assimilations that died out without entering the modern gene pool, but quite a few skeletal remains are said to show clear signs of interbreeding.[33-35]

5. WHAT HAPPENED? Despite rare interbreeding, it is clear that the Neanderthals did not merge genetically with our modern lineage. Therefore, the question arises, why did Neanderthals go extinct? There are many theo-ries. A recent one is based on a volcano that exploded in the Caucasus 40 kya, about the time of arrival of the modern Cro-Magnons, detected as an ash layer in a Russian Neanderthal cave site.[36] The authors argue that this decimated the northern population of Neanderthals, making it easy for the newcomers to occupy that region. It seems like a good idea, but experts comment that it needs more work to persuasively defend.

Another theory is based on sophisticated statistical calculations in population genetics.[37,38] Briefly, those authors show how a long period of rare interbreeding events could lead to extinction (see note with reference). The traditional and still the leading explanation, however, is that the Neanderthals lost out by "competitive exclusion." This, however, is quite vague. There seem to be two attitudes, or general theories, one being that the Neanderthals were intellectually inferior, as discussed above, the other being more consistent with our thesis: that the newcomers had developed a more powerful culture. We already pointed out how the clan-type of marrying society had built-in advantages of cooperation, namely, the clans.

But that cannot be the whole story because careful analysis has shown that the human population density in Europe increased 10-fold between the time of arrival of the Cro-Magnons and the extinction of the Neanderthals.[39] This implies improved exploitation of resources, because it is axiomatic that any population will quickly expand to the maximum capacity of the land, for a given level of efficiency of resource exploitation. Therefore, the Neanderthals were already at maximum population density. What could account for this? Agriculture is out of the question, since it was invented only recently (about 8 kya).

There is no doubt that the Neanderthals' range of habitation shrank southward as the modern humans expanded;[40] and by the way, those authors argue persuasively that another theory, climate change, was *not* the explanation. The vague answer often tendered is "better technology," but exactly what kind of better technology? Anthropologists Shea and Brooks proposed that the modern type may well have had the bow-and-arrow, giving them a great competitive edge.[41] However, the evidence is weak. The invention and diffusion of the bow-and-arrow is an old chestnut, always controversial. Shea later argued that the bow-and-arrow could easily be 100,000 years old,[42] based on the circumstantial finding stone points suited for fitting on arrows.

We may suggest another hypothesis, partly cultural, partly material, to wit, that the new humans brought with them traps and snares for capturing small animals such as rabbits or birds which are otherwise difficult to corner and catch. Only in this way can we explain the observed ten-fold increase in population by the newcomers.[39] That is to say, if the

Neanderthals' meat diet was limited to large game, due to style of hunting or cultural habit, they would have been missing the much larger quantity of potential food present in small game, so this alone could have fostered the population expansion of the Cro-Magnons. One might see here a parallel with a European outpost on Greenland. When the weather went bad, the Europeans starved to death, because they depended on cattle, which all died, and were unable or unwilling to adopt the survival strategies that the native Inuit had showed them.

Oddly enough, the possibility of open hostilities between the newcomers and the Neanderthals is not often discussed, although it seems like a fairly obvious hypothesis (which may be why it is rarely discussed). There is plenty of evidence of violent deaths from those days. For instance, the El Sidron cave in Spain had remains of at least 8 Neanderthals, mostly children and young adults, with evidence of butchering and cannibalism, dated to 43 ky.[43] It is not possible to guess who the attackers were, however. It is known that Neanderthals existed in several regional groups with distinctive race-like morphologic features, such as shape of the face, also supported by mtDNA.[44] They may have been rivalrous. Tooth analysis suggests the victims were starving, perhaps due to absence of mature providers.

Open warfare, albeit sporadic, seems almost certain to have often occurred, if only because good caves were so precious, and rare. Little is known about what kind of houses or tents the Neanderthals may have built. It appears that the Neanderthals exterminated the giant cave bears, which also favored caves. We shall not discuss the famous Chatelperronian cave site, because the dating of its strata is currently controversial. But the general idea is that it held stone tools characteristic of both Neanderthals and Cro-Magnons, which were thought have alternated in its occupation[45] but there is now much doubt because of evidence of mixing of strata.[46,47] It was conjectured on the basis of that site that Neanderthals were engaged in copying the tool styles of the Cro-Magnons,[48] but all such speculation is now doubtful, so we must wait for further clarification from that site and perhaps others.

There are also a large number of theories based on population genetics, such as differing birth rates and time to reach maturity in Neanderthals vs. Cro-Magnons. To mention one of many examples, deLeon et al examined differences in child-bearing between Neanderthals and moderns, and in the closing paragraphs proposes this:

"It could be argued that growing smaller but similarly efficient brains requires less energy investment and might ultimately have led to higher net reproductive rates"[49].

What does he mean by this? Your guess is as good as mine, but is seems to refer to some hypothetical advantage of having a smaller head in birth rates. There seems to be some evidence that Neanderthals matured slightly faster than we do.[50] The whole issue of the relation of brain size to advancement is up in the air, unresolved. There is no doubt that increasing brain size was the hallmark of hominid evolution through *H. erectus*, but beyond that, the whole issue is clouded by almost complete ignorance of how the brain works its magic. We are reminded that modern computers that fit in your pocket are far more powerful than huge ones of a few decades ago. In the opinion of this writer, the many theories based on population genetics need to get their feet on the ground by studying a bit of anthropology, which reveals hundreds of very different societies living cheek-and-jowl in places like New Guinea and North America, directly opposing the theoretical expectation that small advantages possessed by any one of them must eventually lead to it displacing all the others.

That brings us to the bottom line of this article: peaceful coexistence was evidently not possible for Neanderthals and Cro-Magnons, forcing us to conclude that there was a *fundamental* clash of cultures, not merely a different language, or a different religion, or a different system of marrying and kinship, but an almost total absence of one or more of these marks of modern culture in Neanderthals. It was, in a sense, an ideological clash. "Man does not live by bread alone," and this would surely apply also to our paleoancestors, and very likely to all animals. The more recent decimation of primitive societies around the world was caused not only by guns and disease, but by *demoralization:* the Europeans clearly had greater knowledge of the world, and ridiculed the primitives' belief systems, destroying the foundations of their lines, their pride. A similar demoralization might have been at the root of the demise of the Neanderthals.

9.

Digging Deeper

In this chapter, we look at some of the deeper motives and principles underlying the origin, perpetuation, and diversification of the institution of matrimony, and consider it in the context of the great sweep of biological evolution overall.

1. IS IT NATURAL? Although matrimony is clearly a thing of "culture," distinct in our minds from "nature," a wider perspective reveals it to be not so very different from what we see in nature. For example, a new species can arise by preferential inbreeding, avoiding mating with the larger population from which it arose. This amounts to "exogamy," although of course not by any verbal rule, but simply by a "natural preference." But why? What is the basis, the rationale, for such splitting away?

Leaving that for later, we see much the same in any clan-type society: You must marry into another clan of the same society, but not into a completely alien society. Indeed, almost invariably, what we see in humanity today are groups which prefer to marry among themselves. These groups may be ethnic, racial, religious, political, ideological, linguistic, or other. On the other hand, there is a constant and ongoing tendency of such groups to splinter into rivalrous factions, as seen in almost every religion, political system, ethnicity, and so on. These striking parallels with biological species can hardly be pure coincidence.

(Recent genetic studies have revealed the unexpected existence of "cryptic species" among many animals, being those which are visually indistinguishable, yet mate only among themselves. Theoretically, they should not exist, but they do, and may be extremely common. This further underscores the tendency of groups to spontaneously fragment, in the complete absence of any demonstrable materialistic benefit or compulsion.)

On the other hand, other forces are at work to eliminate subgroups and/or to unite them into ever larger wholes. The origin of matrimony as we have portrayed it is a good case-in-point, since it functioned to indissociably weld together formerly independent bands. Again, the history of Christianity is instructive in this regard, as it is seen on the one hand to have unified great numbers, displacing their many prior belief systems, but at the same time, has constantly struggled against disintegrating forces acting to splinter it into endless subordinate denominations, even sub-denominations. Much the same is true of the other major religions. Yet overall, we see the modern world dominated by a comparatively few religions compared with earlier times. The pre-Columbian Amerindians spoke hundreds of different languages, and cleaved to equally or more numerous distinct ideologies and ways of life, whereas modern Americans, occupying the same land but in vastly larger numbers, speak but one language, and are united under a single system of principles and laws.

Thus, matrimony is not a huge departure from the larger world of biology. Birds in particular have invented a marvelous diversity of mating customs, including "monogamy" (life-long pair-bonding). Although birds do not "get married", they are not without their own intricate systems of unspoken rules, as seen, for example, in the habits of many lek-mating birds, some of which exhibit remarkably strict and complicated customs governing their dances, mating displays, and ultimate consummation.

2. DIVORCE & THE COSMOS. Going further, the above-mentioned clear parallels may be extended all the way down the *Scala Natura* and beyond, to the inanimate realm of chemistry and physics. For example, we see in our society a kind of statistical balance, or equilibrium, between married couples and single individuals. This balance would shift if, for example, you dumped a million Russian males here: the fraction of single women would decline. This phenomenon exactly parallels a mathematical topic known as "chemical kinetics" (or steady-state equilibria), describing the tendency of molecular species, A and B, to combine to form AB, and then, after some average time, to break apart, giving back A+B. Indeed, the marriage-divorce parallel is a very good way of introducing the topic of enzyme kinetics to beginners.

What are we to make of this parallel? Just another analogy? Or might this actually reflect the same "natural laws" at work? Alfred J. Lotka, an eminent chemist and physicist of the 1930s, gave numerous other illustrations of how biological systems can be examined by the same mathematical approaches as those otherwise applied to purely chemical (and biochemical) systems. This writer has extended and clarified Lotka's thesis in his book, *The Lotka Hypothesis*, the general conclusion being that the same fundamental principles as drive the rest of the cosmos drive us. We cannot here repeat the logical and fact-based arguments leading to that conclusion but it is sufficiently persuasive to put it beyond reasonable doubt.

The same compelling lines of reasoning also demonstrate that consciousness – our mental lives of desires – can be explained only by supposing that its essence is present in the very matter of which we are composed, and that all of our most fundamental motives, including sexuality, must derive from those which can be inferred from the "laws" of physics and chemistry. A most fundamental principle of this kind is that seen between increasing and decreasing entropy, *i.e.*, disorder, or chaos, *vs.* increasing order, organization. This principle, which has an equivalent formulation in information theory, may be seen in the molecules of the chemist's test-tubes, but equally clearly in the above-mentioned human social and biological phenomena; that is, in the constant splitting of groups and species, *vs.* the elimination of splinter groups and the formation of larger or more embracing wholes.

Likewise, marrying-in *vs.* marrying-out. Every clan-type society – or indeed any unified society, human or other – may be viewed as an incipient "new species," insofar as it excludes or discourages mating with "outsiders." By the same token, our human species is almost certainly the outcome of a large number of variant groups – would-be species of hominids – each of which preferentially mated largely among themselves, only one of which ultimately eclipsed the others. Indeed, this notion has some factual support in the many hominid fossil finds of similar antiquity that have been claimed as distinct species, posing a major puzzle to paleo-anthropologists attempting to establish a clear 1-2-3 phylogeny of our origin. The reality is more like (1a, 1b, 1c) giving rise from 1b to (2a, 2b, 2c) giving rise from 2a to (3a, 3b, 3c)..., and ultimately to ourselves.

Still another parallel is that between the "matrimonial bond" and the *chemical* bond. As pointed out by the great chemist, Linus Pauling, in his classic, *The Chemical Bond*, it is extremely difficult, if not impossible, to define the general "chemical bond," being the various kinds of forces that hold atoms and molecules together. Here again, our train of reasoning leads to the conclusion, which can be supported by technical considerations, that bonds of kinship, marriage, or shared ideologies are direct reflections or manifestations of the self-same forces, at a higher echelon of organization.

3. JUSTICE & RECIPROCITY. In looking for the deeper motives behind the origin of matrimony, it appears that this institution gave voice to a number of new concepts given expression by the advent of true human language. Among these are the linked notions of justice, human equality, and reciprocity. As earlier mentioned, it is safe to assume that in the pre-matrimonial state, offspring within a band were fathered predominately by a comparatively few males, in contrast to the marrying system in which *every* male can have a wife and children. Hence, the marrying system embodies or codifies the concept of human equality, and justice.

The concept of justice, in turn, is bound up with that of fair exchanges. As Levi-Strauss and others have stressed, the idea of bride exchange is near to the essence of the matrimonial system: A father betroths his daughter to another, and in return receives a wife for his son. Indeed, Levi-Strauss was much inspired in his life's work by his countryman and predecessor, Marcel Mauss, famed for his classic, *The Gift*, demonstrating the precision with which gifts and favors are exchanged in human societies, primitive or not. Every trivial favor entails an obligation to repay it in exactly equal measure. Numerous anthropologists have subsequently documented this feature, reciprocity, in many societies.

It has a negative equivalent in terms of insults, offenses, and crimes, each to be repaid – avenged – in kind. For example, adultery is a crime everywhere, perhaps the most common, and many societies allow, or condone, or even expect the cuckolded husband to murder the offender. In some Eskimo communities, murder for reason of adultery was a leading cause of death among men. Years might elapse before the spear was thrust.

The reciprocity of bride exchange and other matters also has its interesting parallels in the inanimate realm of physics and chemistry, where all forces are believed to be mediated by the exchange of sub-atomic particles, and chemical equations are always a "zero-sum game," so to speak, in terms of energy or charge received and released.

4. PROGRESS. Exactly what kind of reason, or motive, or function, or rationality might underlie the tendency of groups to diversity and fragment into splinters? Here we are speaking of the observed variety of marrying systems, viewed as permutations or combinations of the original two-clan system. Although we cannot here paint a full picture, the short answer is the *search for a better way*. The root assumption is that every individual (or group with which he identifies) imagines himself to be perfect, and therefore takes offense when his will or wish is opposed or rejected, or derided, or denied by the others, potentially inspiring him to found a new and different group, more consistent with his wishes, or mien, or logic, or grand vision, along the lines previously exampled.

Therein also lies the mechanism of progress, a.k.a., evolution, and the explanation for diversification. In the human case, it is self-evident that each distinctive group maintains itself as such in the belief that it is better than others in one or more important ways, most notably, in possessing the "real truth" about things, i.e., the belief system underlying the social order. It is clear that allowing doubts about such matters within the group could escalate to internal dissent, factionalism, splintering; therefore all societies attribute the source of their beliefs to super-human entities. Indeed, it is thought that the extreme harshness of rites of initiation in many or most primitive societies functioned preemptively to crush dissent by demanding total unquestioning loyalty and obedience to the reigning order. "Now you are grown, and *one of us*."

The random diversity generated by such splintering amounts to a pool of variants, some of which may occasionally represent a genuine advance, which may or may not emerge as such by virtue of some material advantage, e.g., agriculture, pottery, metals. A non-materialistic example might be Christianity, since it evidently had great intrinsic appeal, enabling it to displace its many fore-runners. Even clearer is the trans-cultural impact of science. The specific features of Christianity which endowed it with such appeal is beyond this discussion, but certainly the notion that God could take a human form is among them, as is the depiction of Jesus as a victim of "this worldly system of things", a fate with which we can all readily identify, and which clearly expresses the "nature" *vs.* "culture" ambivalence. God's laws are the natural laws, if only we could read them correctly.

Thus, the motive for diversification is always the search for a better way – more truthful, more honorable, more profitable, and more efficient. In the great majority of cases, of course, there is no real advantage in the variant forms which arise. And yet, simply by generating such diversity, true advances do sometimes spring forth, to act like an irreversible ratchet, uplifting the species, click by click, over the long haul. The *irreversibility* of true progress, by the way, is yet another key feature of evolution, cultural or other, which has its counterpart in thermodynamics, much as cosmic mentality is implicit in quantum mechanics.

5. MATRIMONY AS PROGRESS. Let us consider again the question, what is behind the great variety of marrying systems world-wide? What is the ultimate underlying motivation? That is to say, we have already indicated that such variations are manifestations of "cosmic laws," and yet, we humans are convinced of some sort of *rationality* to our choices, our decisions. For example, as Lotka pointed out, our economic activities appear to be, statistically speaking, identical in essence to the dumbly impassive "laws of nature," such as thermodynamics. But we know from personal experience that behind every economic transaction lies personal motivation. That is to say, the larger "laws of physics" are in reality nothing more than statistical statements of what actually happens – of what things do – and do not offer any explanations for those doings. Hence, if our economic laws are indeed coextensive with "physical laws," and since we are coextensive with the cosmos from which we sprang, then we must conclude that the real explanation for physical laws is the same as the explanation for our economic laws: motives, desires, will.

These ideas, applied to the general theory of biological evolution, were set forth by this author in another book of this series, *Evolution, Fact & Fantasy: The Psychogenic Hypothesis.* Those same principles must also apply to the human case, including "cultural evolution," which amounts to a short-term snap-shot of much longer-term unfoldings. Thus, as already mentioned, the ultimate motive behind the endless variations on the matrimonial bond (which I just now mis-typed as the "matrimonial *bind*") is to find a better system. And why do we want a better system of mating?

Some additional background explanation of the mechanism of progress is in order here. A particularly clear example of cultural evolution is seen in the march of science. Prior to this march, there were endless theories to explain disease, the heat of the sun, the motion of the planets, and the nature of fire. Those things are now understood beyond reasonable doubt, or at least, sufficiently so for our era, they are settled. The earth goes around the sun, period, end of debate. Any such discovery of a clear-cut factual truth or, what amounts to the same thing, major advance in technology, is an *irreversible* advance, and so defines a step in evolution (progress, as distinct from mere change).

The invention of matrimony was evidently an advance of this kind, insofar as no known human society has ever seen fit to do away with it. The many recent experimental societies that have attempted to do so all ended in failure. But, if it was such an epochal advance, why do we consider it "unnatural," and such an unsatisfactory arrangement? Why have so many permutations of it arisen, all of them presumably in the effort to improve on it?

The answer to that may be seen in the larger world of biology: With few exceptions, the act of sexual mating is fraught with conflict and danger. Why is that? Nobody knows. For that matter, the very existence of sexuality is a major mystery in biology.

6. TOWARDS UNITY? Matrimony was also a significant step in the direction of the afore-mentioned trend towards unification, insofar as it acted to weld together two disparate bands. At the same time, it introduced a compensatory degree of *heterogeneity*, i.e., two (or more) distinct clans living together as one tribe. Today, of course, we see a vastly greater movement in the direction of unification, or at least, of world-wide homogenization. Meanwhile, however, the primitive societies living outside the orb of the mainline civilizations exhibited a huge degree of diversity, such as in the New World, with its hundreds of languages and societies, despite some recent (*ca.* 2000 years) empire-building in Peru, Central America, and perhaps Mississippi. (The northwest coast is an interesting exception, consisting of many distinct and sophisticated societies interacting closely, yet with no evidence of movement toward empire-building or top-down government.)

The question arises, are we headed for world-wide cultural homogeneity? That seems very doubtful. The modern trend towards unification is based on economic inter-dependence (rather than the kinship bonds that unite clans), not so much on shared ideologies. In particular, religions are still very much alive, for the very good reason that they represent the last surviving bastion of spiritual and philosophical debate. Indeed, were it not for Christianity in the western world, philosophy would be all but totally dead.

Accordingly, it is instructive to take heed of some of the more thoughtful and articulate evangelists who warn us of the temptations of Satan, and his many traps and snares. More pointedly, they warn of the temptation to buy into "this worldly system of things," often citing the Book of Revelations. For that matter, the story of Pinocchio – the wooden puppet who yearns to be a real boy, but must first face down the temptations of worldly pleasures – is part-and-parcel of the same philosophico-cultural heritage.

But, why should we not buy into it, with all of its comforts and pleasures of the hour? Why do we not submit and "drown in a sea of love," as one songstress put it? The answer is the same as the explanation for why there were so many different societies (and languages) in the New World: none of them could lay claim to indisputable truth and understanding. The ratchet of progress has yet to turn in this dimension. In other words, we are still waiting for Godot – for the next great irreversible cultural advance, trans-scientific, trans-religious. Everyone knows that this worldly system, including the Priesthood of Academic Empire that support it and forbids original thinking, is corrupt and phony and a pack of lies and half-truths and outmoded traditions. But that next great step has yet to emerge.

And yet, we have faith that it soon will. Hope springs eternal. And until that day dawns, we are stuck. There is no other exit. But when it does, "the trumpets will sound ..." and we shall sing as one voice, "O death, where is thy sting?"

References

References, Ch. 2

1. Staff. ANCIENT JELLYFISH. *Nature* 2007; 450: Nov 15:323. <> (Briefly noted in News column, with reference to source in *PloS*.)
2. Hu Y, et al. LARGE MESOZOIC MAMMAL FED ON YOUNG DINOSAURS. *Nature* 2005; 433: Jan 13:149-152. <> (With commentary by Anne Weil, p116-7, "Living large in the Cretaceous.")
3. Ji Q, Luo ZX, Yuan CX, Tabrum AR. A SWIMMING MAMMALIAFORM FROM THE MIDDLE JURASSIC AND ECOMORPHOLOGICAL DIVERSIFICATION OF EARLY MAMMALS. *Science* 2006; 311: Feb 24:1123-1127. <> (With discussion, p1109-10. Painting of this beaver-like creature is featured on cover.)
4. Luo ZX, Wible JR. A LATE JURASSIC DIGGING MAMMAL AND EARLY MAMMALIAN EVOLUTION. *Science* 2005;308:Apr 1:103-107.
5. Meng J, et al. A MESOZOIC GLIDING MAMMAL FROM NORTHEASTERN CHINA. *Nature* 2006;444:Dec 14:889-893.
6. Luo Z. TRANSFORMATION AND DIVERSIFICATION IN EARLY MAMMAL EVOLUTION [REVIEW]. *Nature* 2007;450:Dec 13:1011-1019.
7. Penny D, Hasegawa M. THE PLATYPUS PUT IN ITS PLACE. *Nature* 1997;387:Jun 5:549-550.<> (Long a puzzle, molecular methods show it separating prior to the marsupials, before K/T boundary.)
8. Ritvo H. THE PLATYPUS AND THE MERMAID AND OTHER FIGMENTS OF THE CLASSIFYING IMAGINATION. Cambridge, MA: Harvard Univ. Press, 1997.<> (Deals with the history of classifying animals, such as the totally unexpected platypus. Reviewed by S. Lyons in *Science*, 279:38, 2 Jan 1998, under headline "Taxonomy Recapitulates Society.")
9. Ji Q, et al. THE EARLIEST KNOWN EUTHERIAN MAMMAL. *Nature* 2002;416:Apr 25:816-822.<> (With perspective by Anne Weil, p798-9.)

10. Kay RF, Ross C, Williams BA. ANTHROPOID ORIGINS. *Science* 1997;275:Feb 7:797-804.<> (The oldest example then known was 37 my, late Eocene of Egypt. The phylogenetic relations shown are based on cladistic analysis, i.e. fossils, not genes. Their analysis is disputed in a later issue, Dec. 19, p2134-6, but authors defend.)

11. Seiffert ER, et al. BASAL ANTHROPOIDS FROM EGYPT AND THE ANTIQUITY OF AFRICA'S HIGHER PRIMATE RADIATION. *Science* 2005;310:Oct 14:300-304.<> (With perspective by Jaeger and Marivaux, p244.)

12. Gebo DL, et al. THE OLDEST KNOWN ANTHROPOID POSTCRANIAL FOSSILS AND THE EARLY EVOLUTION OF HIGHER PRIMATES. *Nature* 2000;404:Mar 16:276-278.<> (Based in new fossil of mouse-size specimen from China, 45 mybp. Report was widely noted, as in *Time* magazine, 3/27/00, p14.)

13. Bloch JL, Boyer DM. GRASPING PRIMATE ORIGINS. *Science* 2002;298:Nov 22:1606-1610.<> (With commentary by E.J. Sargis, p1654-5, "Primate origins nailed.")

14. Janeck JE, et al. MOLECULAR AND GENOMIC DATA IDENTIFY THE CLOSEST LIVING RELATIVE OF PRIMATES. *Science* 2007;318:Nov 2:792-794. (

15. Moya-Sola S, et al. PIEROLAPITHECUS CATALAUNICUS, A NEW MIDDLE MIOCENE GREAT APE FROM SPAIN. *Science* 2004;306:Nov 19:1339-1344.<> (With background, p1273-4.)

16. Suwa G, et al. A NEW SPECIES OF GREAT APE FROM THE LATE MIOCENE EPOCH IN ETHIOPIA. *Nature* 2007;448:Aug 23:921-924.<> (They call it *Chororapithecus ayssinicus*, dated to 10.2 my. For perspective, see News column this issue, p844-5, and *Science* 317:10167.)

17. Simons EL, Ettel PC. GIGANTOPITHECUS. *Scientific American* 1970:Jan:76-85.<>

18. Holden C. APE SEASON COMING UP. *Science* 2005;310:Dec 9:1612.<> (Tells of a documentary film to be made about *Gigantipithecus*, by leading expert, Russell Ciochon. Current estimates place its time from 2 my to 0.3 my, mainly in southern China and northern Vietnam.)

19. McBrearty S, Jablonski NG. FIRST FOSSIL CHIMPANZEE. *Nature* 2005;437:Sep 1:105-108.<> (From Kenya, dated 545 my. By the way, this issue features completion of the chimp genome.)

20. Whitcome KK, Shapiro LJ, Lieberman DE. FETAL LOAD AND THE EVOLUTION OF LUMBAR LORDOSIS IN BIPEDAL HOMININS. *Nature* 2007;450:Dec 13:1075-1078.<> (Shows special adaptation of female vertebrae, to enable upright walking by pregnant females, was present already in early *Australopithecines*.)

21. Spoor F, Wood B, Zonneveld F. IMPLICATIONS OF EARLY HOMINID LABYRINTHINE MORPHOLOGY FOR EVOLUTION OF HUMAN BIPEDAL LOCOMOTION. *Nature* 1994;369:Jun 23:645-648.<> (This report was widely discussed, e.g. in *Sci. American*, Oct., p22, "Standing tall: Inner ear bones provide clues to the emergence of bipedalism".)

22. Brunet M, et al. A NEW HOMINID FROM THE UPPER MIOCENE OF CHAD, CENTRAL AFRICA. *Nature* 2002;418:Jul 11:145-151.<> (This article, with 38 authors, is followed by another, p152-5, about the geology and original environment of the find, dubbed the Toros-Menalla hominid. Informative discussion by Bernard Wood is given on p133-5, with diagram.)

23. Brunet M, et al. NEW MATERIAL OF THE EARLIEST HOMINID FROM THE UPPER MIOCENE OF CHAD. *Nature* 2005;434:Apr 7:752-755.<> (This article is followed by C.P.E. Zollikofer et al, "Virtual cranial reconstruction of *Sahelanthropus tchadensis*," p755-8.)

24. White TD, et al. ASA ISSIE, ARAMIS AND THE ORIGIN OF AUSTRALOPITHECUS. *Nature* 2006;440:Apr 13:883-889.<> (Astounding 31 fossil individuals, dated to 4.12 my, found in Asa Issie area or northeast Ethiopia, west of Aramis, once a woodland. They classify it as *A. anamensis*, the probable immediate ancestor of *A. afarensis*.)

25. White T. EARLY HOMINIDS: DIVERSITY OR DISTORTION? *Science* 2003;299:Mar 28:1994-1996.<> (He decries the trend of naming endless fossil hominids as species, now numbering at least 20, hinting that political correctness about "diversity" may be involved, and citing instructive examples from the past: a great many types of pig fossils were reduced when it was realized how geology could distort them. He also points out that wide differences exist in any giving population. However, he offers no simplifying scheme of his own, or at least, not until 2006 due to his own fossil finds.)

26. Johanson D, Edgar B. FROM LUCY TO LANGUAGE. New York: Simon and Schuster, 2006.<> (Gorgeous book featuring full-size color photographs of perhaps 200 of the most important fossil specimens, each explained in some detail.)

27. Gibbons A. BREAKTHROUGH OF THE YEAR: ARDIPITHECUS RAMIDUS. *Science* 2009;326:Dec18:1598-1599.<> This brief article refers to the issue of Oct 2, pg 60-106, for detailed reports.

28. Johanson DC, Edey MA. LUCY. THE BEGINNINGS OF HUMANKIND. New York: Simon & Schuster, 1981.

29. Hay RL, Leakey MD. THE FOSSIL FOOTPRINTS OF LAETOLI. *Scientific American* 1982:Feb:50-57.<>

30. Sponheimer M, et al. ISOTOPIC EVIDENCE FOR DIETARY VARIABILITY IN THE EARLY HOMININ PARANTHROPUS ROBUSTUS. *Science* 2006;314:Nov 10:980-982.<> (Photos and commentary p930 by Ambrose.)

31. Staff. BIGTOOTH LIKED ITS FOOD CRUNCHY. *New Scientist* 2005:Oct 29:15.<> (Highlights of 65th annual meeting of *Society of Vertebrate Paleontology*. One paper presented evidence that stout teeth such as in *Paranthropus* are similar to those of otters and others which feed on shellfish.)

32. Goodall J. TOOL-USING AND AIMED THROWING IN A COMMUNITY OF FREE-LIVING CHIMPANZEES. *Nature* 1964;201:1264-1266.

33. Bower B. CHIMPS SPREAD OUT THEIR TOOLS. *Science News* 2006;170:Sep 2:157.<> (Note on work by B.J. Morgan and colleagues appearing in *Current Biology*, issue Aug. 20, showing that a local innovation - opening nuts by cracking between two stones - had spread some distance to other groups, crossing a river. Related reports are discussed in more depth and p154-6.)

34. Mulcahy NJ, Call J. APES SAVE TOOLS FOR FUTURE USE. *Science* 2006;312:May 19:1038-1049.<> (See p1006-7 for commentary, by T. Suddenhorf, with the ambitious title, "Foresight and evolution of the human mind." Widely discussed elsewhere, e.g. by E. Jaffe in *Science News* 170:154-7, Sep. 2.)

35. Gibbons A. SPEAR-WIELDING CHIMPS SEEN HUNTING BUSH BABIES. *Science* 2007;315:Feb 23:1063.<> (Chimp females and youngsters, who are apparently not allowed to hunt green monkeys with males, are seen to thrust sticks, sharpened with their incisors, into holes in trees where bush babies sleep in day time. Also noted in *New Scientist*, Mar. 13 p16, telling that chimps are observed to seek shelter in caves.)

36. Whiten A, Horner V, deWaal FBM. CONFORMITY TO CULTURAL NORMS OF TOOL USE IN CHIMPANZEES. *Nature* 2005;437:Sep 29:737-740.

37. Vogel G. CHIMPS IN THE WILD SHOW STIRRINGS OF CULTURE. *Science* 1999;284:Jun 25:2070-2073.<> (Good survey of the evidence and arguments as of then. Followed by a piece titled, "Are our primate cousins 'conscious'?" p2073-9.)

38. Whiten A. THE SECOND INHERITANCE SYSTEM OF CHIMPANZEES AND HUMANS. *Nature* 2005;437:Sep 1:52-55.<> (In special issue announcing the chimp genome, with several other relevant articles, such as by the primatologist, Franz deWaal, p56-9. For letter critical of the term "culture" applied to apes, see issue of Nov. 24 p422; and Whiten's response, issue of Dec. 22/29 p1078.)

39. Holmes B. CULTURED CHIMPANZEES GET THE MESSAGE. *New Scientist* 2006:Sep 2:11.<> (Note on work appearing in *Proc. Nat'l. Acad. Sci.* demonstrating that chimps can accurately transmit new behavioral strategies from one to the next and so on.)

40. Grant B. DO CHIMPS HAVE CULTURE? *The Scientist* 2007:Aug:29-35.<> (Survey of recent field and laboratory studies bearing on the question.)

41. Cohen J. THE WORLD THROUGH A CHIMP'S EYES. *Science* 2007;316:Apr 6:44-45.<> (Highlights of a symposium, "The mind of the chimpanzee," attended by many leaders, including Jane Goodall, the grand dame of the field, who remarks that in the 1960's it was verboten to even speculate that chimps could have minds and feelings similar to our own.)

42. Blumenschine RJ, Cavallo JA. SCAVENGING AND HUMAN EVOLUTION. *Sci Amer* 1992:Oct:90-96.<>

43. Zimmer C. FASTER THAN A HYENA? RUNNING MAY MAKE HUMANS SPECIAL. *Science* 2004;306:Nov 19:1283.

44. Thorpe SKS, Holder RL, Crompton RH. ORIGIN OF HUMAN BIPEDALISM AS AN ADAPTATION FOR LOCOMOTION ON FLEXIBLE BRANCHES. *Science* 2007;316:Jun 1:1328-1331. (Perspective by O'Higgins and Elton p1292-4. Critiques of this theory appear in the issue of Nov. 16, 318:1065-6.)

45. Bower B. RED APE STROLL. *Science News* 2007;172:Aug 4:72-73.

46. Richmond BG, Strait DS. EVIDENCE THAT HUMANS EVOLVED FROM A KNUCKLE-WALKING ANCESTOR. *Nature* 2000;404:Mar 23:382-385.<> (With perspective p339 by M. Collard and L.C. Aiello.)

47. Lee RB, DeVore I. MAN THE HUNTER. Hawthorne, NY: Aldine Co., 1968.<> (Proceedings of a symposium at the Univ. of Chicago organized by Sol Tax, in honor of Claude Levi-Strauss. Focus on hunter-gatherers, marriage, and kinship, several on Australia, etc.)

48. Washburn SL. THE EVOLUTION OF MAN. *Scientific American* 1978:Sep:194-207.<>

49. Nitecki MH, Nitecki DV. THE EVOLUTION OF HUMAN HUNTING. New York: Plenum Press, 1987.<> (Proceedings of a symposium, reviewed in *Science* 243:241, by J.D. Speth.)

50. Leakey LSB. OLDUVAI GORGE. *Scientific American* 1954:Jan:66-71.<>

51. Walker A, Leakey R. THE HOMINIDS OF EAST TURKANA. *Scientific American* 1978:Aug:54-66.<>

52. Walker A, Leakey R. THE NARIOKOTOME HOMO ERECTUS SKELETON. Cambridge, MA: Harvard Univ. Press, 1993.<> (Among the best fossil hominids. Found in northern Kenya in 1984, dated to 1.53 my, was a boy 11-15 years old of size indicating that the adult stood 1.8 meters (5'9") tall. Often called WT 15000, its number in the Kenya museum. Reviewed in *Science* 265:418.)

53. Weiner S, Xu Q, Goldberg P, Liu J, Bar-Yosef O. EVIDENCE FOR THE USE OF FIRE AT ZHOUKOUDIAN, CHINA. *Science* 1998;281:Jul 10:251-253.<> (With discussion, p165-6, "Geological analysis damps ancient Chinese fires." Also, book review on history of fire, p180-1.)

54. Goren-Inbar N, et al. EVIDENCE OF HOMININ CONTROL OF FIRE AT GESHER BENOT YA'AQOV, ISRAEL. *Science* 2004;304:Apr 30:725-727.<> (With commentary, p663-4.)

55. Gibbons A. FOOD FOR THOUGHT. *Science* 2007;316:Jun 15:1558-1560.<> (Highlights of conference, "Primatology meets paleo-anthropology," on origin and significance of cooking.)

56. Holmes B. TUBERS, THE BRAIN FOOD OF CHOICE? *New Scientist* 2007:Sep 15:18.<> (Source cited is *Nature Genetics*. Also discussed in *Nature* 449:155, and *Science News* 172:173. Among other things, they find that ethnic groups which rely heavily on starch have more copies of the amylase gene, and secrete more of this enzyme in their saliva. Interesting though that is, it is not clear how that finding bears on their thesis.)

57. Isaac G. THE FOOD-SHARING BEHAVIOR OF PROTOHUMAN HOMINIDS. *Scientific American* 1978:Apr:90-109.<>

58. Parker I. SWINGERS. *New Yorker* 2007:Jul 30.<> (Balanced account of chimpanzee disposition, especially bonobos, as reported by various eminent observers, whose views and conclusions are often quite different, even contradictory. Several relevant books and articles are cited.)

59. Bower B. EVOLUTIONARY BACK STORY. *Science News* 2006;169:May 6:275-276.<> (On work reported by Marc M. Hauser and colleagues to the *Paleoanthropology Society* meeting in San Juan, that analysis of oldest known vertebrae of Homo, from 1.8 mybp *H. erectus* found 2005 in Dmansi, Georgia, shows it is within normal modern human range, and sufficient support for respiratory muscles controlling speech, in contrast to previously oldest erectus, found 1984 in Kenya dated 1.6 mybp [which some call H. ergaster], had smaller chimp-like vertebrae. But others such as Bruce Latimore think the Kenya specimen had malformed or stunted bones, possibly malnourished.)

60. Lieberman P. VOICE IN THE WILDERNESS: HOW HUMANS ACQUIRED THE POWER OF SPEECH. *The Sciences* 1988:Jul/Aug:23-29.<>

61. Lieberman P. HUMAN LANGUAGE AND OUR REPTILIAN BRAIN. Cambridge, MA: Harvard Univ. Press, 2000.<> (Subtitle, "The subcortical bases of speech, syntax and thought." Regarding Neanderthals, discusses his hypothesis, coauthored with Crelin, in light of new findings, p135-150, and see Note 5, p176-7.)

References, Ch. 3.

1. Morgan LH. LEAGUE OF THE HO-DE-NO-SAU-NEE, OR IROQUOIS. Rochester, N.Y.: Sage and Broa, 1851. (It appears that Morgan was unaware of Lafitau's earlier work, which he independently rediscovered.)

2. Morgan LH. SYSTEMS OF CONSANGUINITY AND AFFINITY OF THE HUMAN FAMILY. Washington, D.C.: Smithsonian Institution, 1870.<> (Original, 1877.)

3. Bachofen JJ. DAS MUTTERRECT [THE MOTHER RIGHT]. Basel, Suisse: Benno Schwabe, 1861.

4. Graves R. THE WHITE GODDESS. NY: Octagon Books / Farrar, Strauss & Giroux, 1978.

5. Davis EG. THE FIRST SEX. Baltimore, MD.: Penguin Books, 1972.

6. Maine SHS. ANCIENT LAW. London: J. Murray, 1861.<> (Subtitle, *Its connection with the early history of society and its relations to modern ideas*. Later works continued to the 1880's.)

7. Farb P. MAN'S RISE TO CIVILIZATION AS SHOWN BY THE INDIANS OF NORTH AMERICA FROM PRIMEVAL TIMES TO THE COMING OF THE INDUSTRIAL STATE. New York: E.P. Dutton and Co., 1968.<> (From the Foreword by E. Service: "It is the best general book about American Indians that I have ever read ... it departs strikingly from the standard works, such as Wissler's classic in anthropology, *The American Indian*, by being readable. Peter Farb studied his subject hard for years, but he writes like a breeze ..." All that is true but readers should be aware that the purpose of the book is also to defend White's theory.)

8. Macleod M. WHEN FEMALE CHIMPS TURN TO INFANTICIDE. *New Scientist* 2007: May 19:19.<> (Note on field observations by Simon Townsend and colleagues appearing in *Current Biology* v17:R356.)

9. Morgan LH. ANCIENT SOCIETY. N.Y. (1963): World Publishing / Meridian Books, 1877.

10. Freud S. TOTEM AND TABOO. NY, NY: Moffat, Yard, 1918.<> (Translated by A.A. Brill.)

11. Levi-Strauss C. TOTEMISM. Boston MA: Beacon Press, 1963.

12. White LA. THE EVOLUTION OF CULTURE. New York: McGraw Hill Inc. (paperback), 1959.

13. Johanson DC, Edey MA. LUCY. THE BEGINNINGS OF HUMANKIND. New York: Simon & Schuster, 1981.

14. Lovejoy CO. THE ORIGIN OF MAN. *Science* 1981;211:Jan 21:341-350.

15. Lovejoy CO. EVOLUTION OF HUMAN WALKING. *Scientific American* 1988:Nov:118-126.

16. Wilson EO. SOCIOBIOLOGY: THE NEW SYNTHESIS. N.Y.: Cambridge Univ. Press, 1975.

17. Pilbeam D. THE EVOLUTION OF MAN. New York: Funk & Wagnalls Press, 1970.

18. Hagman M. MORE QUESTIONS ABOUT THE PROVIDER'S ROLE. *Science* 1999;283:Feb 5:777.<> (Highlights from the Anthropology division of the annual convention of

AAAS. Major theme seems to be new doubts about theories such as Lovejoy's which stress the importance of the father's role as provider. Interesting reports. Interesting.)

19. Malinowski B. Magic, Science and Religion. New York: Doubleday (paperback reissue), 1955.

20. Malinowski B. Argonauts of the Western Pacific. New York: E.P. Dutton; originally 1922, 1961.<> (His evidence for ignorance of paternity consisted of several observations. First, belief by women that they got pregnant by sitting in certain seaside caves, second, the complete absence of suspicion of infidelity if a woman became pregnant in her husband's absence, and other arguments which I do not recall. This claim was later doubted but not convincingly. The Melanesians had been isolated for a very long time, probably as long or longer than Australians, and thus may reflect a relatively primitive state.)

21. Wilson PJ. Man, the Promising Primate. New Haven, CT: Yale Univ. Press, 1980.

22. Sahlins MD, Service ER. Evolution and Culture. Detroit, MI: Univ. of Michigan Press, 1966.

23. Meggers BJ. The law of cultural evolution as a practical research tool. In: Dole GE, Carneiro RL, eds. Essays in the Science of Culture in Honor of Leslie A. White. N.Y., N.Y.: Thomas Y. Crowell Co., 1960:302-316.

24. Gregg SA. Between Bands and States Occasional Paper #9, Center for Archeological Investigations. Chicago, IL: Southern Illinois Univ. Press, 1991. (Reviewed in *Amer. Scientist,* 81:394-5, Jul/Aug 1993, by Paul D. Welch, who notes continuing interest "in the 30 years since Ellman Service proposed the band – tribe – chiefdom – state evolutionary typology." The 20 papers in this book address the question, "how can we conceptualize small-scale, acephalous [headless] societies in nonstate or prestate contexts?" Perspective is mainly archeological "but also includes chapters that deal with ethnographic cases." At least six chapters focus on Amerindians.)

25. Suttles W. Coping with abundance: subsistence on the Northwest Coast [Ch. 6]. In: Lee RB, DeVore I, eds. Man the Hunter. Hawthorne, NY: Aldine Co., 1968:56-68.<> (Proceedings of a symposium at the Univ. of Chicago organized by Sol Tax, in honor of Claude Levi-Strauss. Focus on hunter-gatherers, marriage, and kinship, several on Australia.)

26. Pilling AS. Southeastern Australia: levels of social organization [Ch. 16]. In: Lee RB, DeVore I, eds. Man the Hunter. Hawthorne, NY: Aldine Co., 1968:138-146.<> (Casts doubt on the theory of Sahlins and White. Among other things, shows that reports of chiefs or kings in Australia were erroneous, due in part to the practice of white landowners or governors of appointing local chiefs, headmen or kings.)

27. White L. The Science of Culture. New York: Farrar, Strauss and Co,, 1949.<> (Includes essay on public education, showing that American attitudes towards science are deeply ingrained in our cultural fabric and that public education cannot be improved without changes in the culture.)

28. Darwin C. The Descent of Man and Selection in Relation to Sex. London: John Murray (2nd ed'n), 1874.

References, Ch. 4.

1. White LA. THE EVOLUTION OF CULTURE. New York: McGraw Hill Inc. (paperback), 1959.

2. Farb P. MAN'S RISE TO CIVILIZATION AS SHOWN BY THE INDIANS OF NORTH AMERICA FROM PRIMEVAL TIMES TO THE COMING OF THE INDUSTRIAL STATE. New York: E.P. Dutton and Co., 1968.<> (From the Foreword by E. Service: "It is the best general book about American Indians that I have ever read ... it departs strikingly from the standard works, such as Wissler's classic in anthropology, *The American Indian,* by being readable. Peter Farb studied his subject hard for years, but he writes like a breeze ..." All that is true, but readers should be aware that the purpose of the book is also to defend White's theory.)

3. Levi-Strauss C. TOTEMISM. Boston MA: Beacon Press, 1963.

4. Freud S. TOTEM AND TABOO. NY, NY: Moffat, Yard, 1918.<> (Translated by A.A. Brill. Republished several times. See also his *Future of an Illusion.*)

5. Waters F. BOOK OF THE HOPI. New York: Ballantine Books (originally Viking, 1963), 1969.

6. Hiatt LR. GIDJINGALI MARRIAGE ARRANGEMENTS [CH. 18]. In: Lee RB, DeVore I, eds. Man the Hunter. Hawthorne, NY: Aldine Co., 1968:165-175.<> (But see also Ch. 22 for discussion of Ch. 18-21 [Part IV, *Marriage and Models in Australia*], including some barbed exchanges and sharp rebuttals, such as from Levi-Strauss himself, in whose honor this symposium was held.)

7. Meggitt MJ. "MARRIAGE CLASSES" AND DEMOGRAPHY IN CENTRAL AUSTRALIA [CH. 19]. In: Lee RB, DeVore I, eds. Man the Hunter. Hawthorne, NY: Aldine Co., 1968:176-185.<> (But see also Ch. 22 for discussion of Ch. 18-21 [Part IV, *Marriage and Models in Australia*], including some barbed exchanges and sharp rebuttals, such as from Levi-Strauss himself, in whose honor this symposium was held.)

8. Service ER. SOCIOCENTRIC RELATIONSHIP TERMS AND THE AUSTRALIAN CLAN SYSTEM. In: Dole GE, Carneiro RL, eds. Essays in the Science of Culture in Honor of Leslie A. White. N.Y., N.Y.: Thomas Y. Crowell Co., 1960:416-436.<> (Reviews alternative concepts of the function or purpose of the clan systems in Australia, as set forth by the leading specialists, apart from their role in regulating marriage. However, does not discuss their function of reflecting native cosmology.)

9. Levi-Strauss C. THE SAVAGE MIND. Chicago, IL: Univ. of Chicago Press, 1966.<> (Original French, *Le Pensèe Sauvage*; 1962, Librairie Plon, Plaris. First English edition, 1966, George Weidenfeld and Nocholson Ltd., London. It is said that this title plays on Boas' book, *The Mind of Primitive Man* (1938, paperback 1963, The Free Press / MacMillan), and that the French title also means, in the vernacular, *The Wild Pansy*, a fact of emergent significance in his later works. The word *pensèe* is difficult to translate – thoughts? mind? ideas? – as seen also in Blaise Pascal's famous *Pensèes*, usually left untranslated in English editions.)

10. Morgan LH. SYSTEMS OF CONSANGUINITY AND AFFINITY OF THE HUMAN FAMILY. Washington, D.C.: Smithsonian Institution, 1870.<> (Original, 1877.)

11. Eggan F. LEWIS H. MORGAN IN KINSHIP PERSPECTIVE. In: Dole GE, Carneiro RL, eds. Essays in the Science of Culture in Honor of Leslie A. White. N.Y., N.Y.: Thomas Y. Crowell Co., 1960:179-201.<> (By the way, Eggan notes that the French Jesuit Missionary, Lafitau, had made the same observations as Morgan about the Iroquois and some others, as early as 1724. This helps confirm that Morgan's findings a century later were not yet distorted by contact with white people.)

12. Powdermaker H. STRANGER AND FRIEND: THE WAY OF AN ANTHROPOLOGIST. New York: W. W. Norton & Co., Inc., 1966.<> (Her memoirs, for a popular audience. Begins with her work with Melanesians in the village of Lesu on New Ireland, a Pacific island, beginning 1929, with anecdotes about here teacher, Malinowski, and other luminaries such as Radcliffe-Brown. Then moves to her other studies, such as of black-white relations in a southern American town.)

13. Levi-Strauss C. THE ELEMENTARY STRUCTURES OF KINSHIP. Boston, MA: Beacon Press, 1969.<> (Original French: *Les Structures Elementaires de la Parente*. Original French edition, 1949; Presses Universitaires de France.)

14. Murdock GP. SOCIAL STRUCTURE. New York: Macmillan, 1949.

15. Dole GE. THE CLASSIFICATION OF YANKEE NOMENCLATURE IN THE LIGHT OF EVOLUTION OF KINSHIP. In: Dole GE, Carneiro RL, eds. Essays in the Science of Culture in Honor of Leslie A. White. N.Y., N.Y.: Thomas Y. Crowell Co., 1960:162-178.<> (Main aim is to dispute Murdock's influential work concluding that modern American (Yankee) kinship is essentially the same as Eskimo and Andaman, disproving the assumption of evolutionists like White that kinship systems are tied to levels of technology and social complexity. Analyses and diagrams are very clear and useful, and author may be correct in limited degree but impression is one of nit-picking.)

16. Goodenough WH. YANKEE KINSHIP TERMINOLOGY: A PROBLEM IN COMPONENTIAL ANALYSIS. In: Tyler SA, ed. Cognitive Anthropology. New York: Holt, Rinehard and Winston, Inc,, 1969:255-287.<> (Interesting, somewhat novel approach. For critique, see p288 by D.M. Schneider. See also p369-396, *Cognitive aspects of English kin terms*, by A. Kimball Romney and R.G. D'Andrade.)

17. Lee RB, DeVore I. MAN THE HUNTER. Hawthorne, NY: Aldine Co., 1968.<> (Proceedings of a symposium at the Univ. of Chicago organized by Sol Tax, in honor of Claude Levi-Strauss. Focus on hunter-gatherers, marriage, and kinship, several on Australia, etc.)

18. Turnbull CB. THE FOREST PEOPLE. New York: Simon and Schuster (Touchstone paperback), 1962.<> (Charming account of the ethnologist's life among the BaMbuti Pygmies of the Ituri Forest of Central Africa. The book is dedicated not to his wife, mom or teacher but "To Kenge," his Pygmie friend, "for whom the forest was Mother and Father, Lover and Friend; and who showed me something of the love that all his people share in a world that is still kind and good, and without evil." It is worth noting that this sentiment is widely felt towards other hunting and gathering peoples such as the Kung Bushmen and Australians.)

19. Pilling AS. SOUTHEASTERN AUSTRALIA: LEVELS OF SOCIAL ORGANIZATION [CH. 16]. In: Lee RB, DeVore I, eds. Man the Hunter. Hawthorne, NY: Aldine Co., 1968:138-146.<> (Casts doubt on the theory of Sahlins and White. Among other things, shows

that reports of chiefs or kings in Australia were erroneous, due in part to the practice of white landowners or governors of appointing local chiefs, headmen or kings.)

20. Turnbull CM. THE IMPORTANCE OF FLUX IN TWO HUNTING SOCIETIES [CH. 15]. In: Lee RB, DeVore I, eds. Man the Hunter. Hawthorne, NY: Aldine Co., 1968:132-137.

21. Balikci A. THE NETSILIK ESKIMOS: ADAPTIVE PROCESSES [CH. 8]. In: Lee RB, DeVore I, eds. Man the Hunter. Hawthorne, NY: Aldine Co., 1968:78-83.<> (See also Ch. 12, *The diversity of Eskimo societies*, by David Damas.)

22. Mead M. SEX AND TEMPERAMENT IN THREE PRIMITIVE SOCIETIES. N.Y.: Mentor Books / New American Library. (First ed'n 1935, Morrow), 1950.<> (See also, *Coming of Age in Samoa,* etc.)

23. Freeman D. MARGARET MEAD AND SAMOA: THE MAKING AND UNMAKING OF AN ANTHROPOLOGICAL MYTH. 1983.<> (This book made headlines, such as in a January issue of the *New York Times:* "New Samoa book challenges Margaret Mead's conclusions.")

24. Harris M. MARGARET AND THE GIANT-KILLER. *The Sciences* 1983;23:Jul/Aug:18-21`.<> (Further defense of Mead against Freeman's critique, and he takes the opportunity to also cogently discuss the issue of nature vs. nurture, since Freeman makes much o the "absolute cultural determinism" taught under Boas, which Harris shows to be wrong. He capably demolishes much of Freeman's position. By the way, Mead was a student of Ruth Benedict, who in turn was a student of Boas. Also by the way, Harris remarks that it is the four-field approach that distinguishes anthropology from sociology in the U.S., not "the study of so-called primitive people." His conclusion on the Mead affair: "Perhaps Mead overemphasized the bright and amiable side of Samoan life; but perhaps Freeman has overemphasized the dark and aggressive side.")

25. Holmes LD. SOUTH SEAS SQUALL. *The Sciences* 1983;23:Jul/Aug:14-17.<> (Critical review and rejoinder to "Derek Freeman's long-nurtured, ill-natured attack on Margaret Mead." Holmes is well qualified to mediate since he had personally restudied the people and places that Mead had written about nearly 30 years before. All parties admit that Mead had some degree of bias or interpretive slant but not more than in most such reports of her day. Holmes confirms most of her observations and explains the few significant discrepancies with Freeman by such factors as the different generation of observation.).

26. Horstman LL, Horstman CL. THE LOTKA HYPOTHESIS, BOOK I: ELEMENTS OF CONSCIOUSNESS. New York: Vantage Press, 2006.

27. Tyler SA. COGNITIVE ANTHROPOLOGY. New York: Holt, Rinehart and Winston, Inc., 1969.<> (A collection of about two-dozen essays in five sections. Several are key papers reprinted from other sources. Leaders include Floyd G. Lounsbury, Ward H. Goodenough, A. Kimball Romney, Roy G. D'Andrade, Mary B. Black, Charles O. Flake, Dell H. Hymes, and others. Many of the papers presuppose a strong background, e.g. in kinship, linguistics. Several of these papers are excellent and fascinating, but not because of the touted "formal methods" such as "componential analysis." Some of the papers strike me as just jargon-laden hot-air.)

28. Harris M. THE RISE OF ANTHROPOLOGICAL THEORY. N.Y.: Thomas Y. Crowell & Co. (1st edn), 1968.<> (Updated edition appeared 2001 (Rowan & Littlefield) with

introduction on this history of the book, now a classic, by M.L. Margolis. The book assumes that the reader is already familiar with the topics and authors discussed. Plenty of downside but remains a good survey and a rich source of key literature.)

29. Johnson A. THE DEATH OF ETHNOGRAPHY: HAS ANTHROPOLOGY BETRAYED ITS MISSION? *The Sciences* 1987:Mar/Apr:23-30.

30. Editors. CLIFFORD GEERTZ BY HIS COLLEAGUES. Chicago: Univ. of Chicago Press, 2005.<> (Briefly noted in *New Scientist,* Mar 26, 2005, p55.)

31. Kroeber AL, Kluckohn C. CULTURE: A CRITICAL REVIEW OF CONCEPTS AND DEFINITIONS. N.Y.: Random House (Vintage, paperback), 1952.<> (Originally in vol XLVII of the *Papers of the Peabody Museum of American Archaeology and Ethnology,* Harvard University. It is said that it was written mainly by Kluckohn, then edited by the more famous and prestigious Kroeber.)

32. Cafagna AC. A FORMAL ANALYSIS OF DEFINITIONS OF 'CULTURE'. In: Dole GE, Carneiro RL, eds. Essays in the Science of Culture in Honor of Leslie A. White. N.Y., N.Y.: Thomas Y. Crowell Co., 1960:111-132.<> (Gives short review of various definitions and their weaknesses, and ideas about definitions generally, but reaches no conclusion. Aim seems to be to clarify and improve White's definition in terms of "symbolling," i.e., White viewed *language* as the central feature of culture.)

33. Wrangham RW, McGrew WC, deWaal FBM, Heltne PG. CHIMPANZEE CULTURES. Cambridge, MA: Harvard University Press, 1997.

34. Mulcahy NJ, Call J. APES SAVE TOOLS FOR FUTURE USE. *Science* 2006;312:May 19:1038-1049.<> (See p1006-7 for commentary, by T. Suddenhorf, with the ambitious title, "Foresight and evolution of the human mind". Widely discussed elsewhere, e.g. by E. Jaffe in *Science News* 170:154-7, Sep. 2.)

35. Gibbons A. SPEAR-WIELDING CHIMPS SEEN HUNTING BUSH BABIES. *Science* 2007;315:Feb 23:1063.<> (Chimp females and youngsters, who are apparently not allowed to hunt green monkeys with males, are seen to thrust sticks, sharpened with their incisors, into holes in trees where bush babies sleep in day time. Also noted in *New Scientist*, Mar. 13 p16, telling that chimps are observed to seek shelter in caves.)

36. Holmes B. STONE AGE CHIMPS WERE HANDY WITH A HAMMER. *New Scientist* 2007:Feb 17:15.<> (Cites work appearing in Proc. Nat'l. Acad. Sci. on evidence that chimps had a "stone age" in which they crafted stone tools. However, experts consulted for this article and others on this report [*Sci. News,* 171:88-9 and 171:99] are extremely skeptical, as am I, although all agree that chimps use stones to hammer open tough nuts.)

37. Grant B. DO CHIMPS HAVE CULTURE? *The Scientist* 2007:Aug:29-35.<> (Survey of recent field and laboratory studies bearing on the question.)

38. Holmes B. CULTURED CHIMPANZEES GET THE MESSAGE. *New Scientist* 2006:Sep 2:11.<> (Note on work appearing in *Proc. Nat'l. Acad. Sci.* demonstrating that chimps can accurately transmit new behavioral strategies from one to the next and so on.)

39. Bower B. CHIMPS SPREAD OUT THEIR TOOLS. *Science News* 2006;170:Sep 2:157.<> (Note on work by B.J. Morgan and colleagues appearing in *Current Biology*, issue Aug. 20, showing that a local innovation - opening nuts by cracking between two

stones - had spread some distance to other groups, crossing a river. Related reports are discussed in more depth and p154-6.)

40. Whiten A. THE SECOND INHERITANCE SYSTEM OF CHIMPANZEES AND HUMANS. *Nature* 2005;437:Sep 1:52-55.<> (In special issue announcing the chimp genome, with several other relevant articles, such as by the primatologist, Franz deWaal, p56-9. For letter critical of the term "culture" applied to apes, see issue of Nov. 24 p422; and Whiten's response, issue of Dec. 22/29 p1078.)

41. Vogel G. CHIMPS IN THE WILD SHOW STIRRINGS OF CULTURE. *Science* 1999;284:Jun 25:2070-2073.<> (Good survey of the evidence and arguments as of then. Followed by a piece titled, "Are our primate cousins 'conscious'?" p2073-9.)

42. Cohen J. THE WORLD THROUGH A CHIMP'S EYES. *Science* 2007;316:Apr 6:44-45.<> (Highlights of a symposium, "The mind of the chimpanzee," attended by many leaders, including Jane Goodall, the grand dame of the field, who remarks that in the 1960's it was verboten to even speculate that chimps could have minds and feelings similar to our own.)

43. Pennisi E. SOCIAL ANIMALS PROVE THEIR SMARTS. *Science* 2006;312:Jun 23:1734-1738.<> (Survey of recent studies, not limited to higher primates but including birds, dogs, hyenas, etc. I wish they had included squirrels, who are also very clever.)

44. Staff. SPIDER MONKEYS GO ON THE WARPATH. *New Scientist* 2006:May 27:14.<> (Note on report in recent *Amer. J. Phys. Anthropol.*, by F. Aureli and colleagues, reporting that spider monkey males carry out raids and sneak attacks on neighboring groups, behavior previously seen in chimps.)

45. Spinney L. WHAT ONLY A CHIMP KNOWS. *New Scientist* 2006:Jun 10:48-49.<> (Interview with eminent primatologist, Tetsuro Matsuzawa, a chimp expert, who mentions, among other things, that chimps make use of some 200 of the 600 forest plants, and "have a botanists memory for them.")

46. Goodall J. TOOL-USING AND AIMED THROWING IN A COMMUNITY OF FREE-LIVING CHIMPANZEES. *Nature* 1964;201:1264-1266.

References, Ch. 5.

1. Berndt RM. Excess and Restraint. Chicago, IL: Univ. Chicago Press, 1962.<> (See Ch. 12, "Warfare." Book is detailed study of several tribes of New Guinea central highlands.)

2. Tylor EB. On a method of investigating the development of institutions: Applied to laws of marriage and descent. *J Royal Anthrop Institute* 1889;18:245-269.<> (As given by Marvin Harris.)

3. White LA. The Evolution of Culture. New York: McGraw Hill Inc. (paperback), 1959.

References, Ch. 6.

1. Prat S, Brugal JP, et al. First occurrence of early Homo in the Nachukui Formation (West Turkana, Kenya) at 2.3-2.4 Myr. *J Hum Evol* 2005;49:2:230-240.

2. Vekua A, et al. A new skull of early Homo from Dmanisi, Georgia. *Science* 2002;297:Jul 5:85-89.<> (With perspective and comments from several experts, p26-7, "Were 'little people' the first to venture out of Africa?" This new skull and mandible, dated to 1.75 my, is the third and best found at this site, but is smaller, perhaps an adolescent. They classify it as early *H. erectus* but stress that it has several features of *H. habilus*.)

3. Spoor F, Leakey MG, Gathogo PN, Brown FH, Anton SC, McDougall I, Kiarie C, Manth FK, Leakey LN. Implications of new early *Homo* fossils from Ileret, east of Lake Turkana, Kenya. *Nature* 2007;448:Aug 9:688-691.<> (Find is an unusually small *erectus* skull (calvaria) of unexpectedly late date, and a partial mandible assigned to *H. habilus*. The original habilus found by Louis Leakey in 1960 is dated 2.0 my, and is here said to have persisted well beyond the first *H. erectus*. Also discussed is suggestive evidence that these early hominids were sexually dimorphic, i.e., males considerably larger than females. But this calvaria was likely a sub adult. Also, doubts are raised about clarity of distinction of Asian *H. erectus*.)

4. Holmes B. Fossil jaw speaks of long-lingering ancestor. *New Scientist* 2007:Aug 11:12.<> (Comments on article in *Nature* by Spoor et al claiming H. habilus persisted beyond H. erectus but some authorities interviewed were very skeptical, such as Tim White: "I'll be charitable ... to say it'll be controversial." Also discussed in Science 317:733, similarly. But all agree that it is an interesting and important find. J. Schwartz is quoted saying, "the more this stuff comes out, the more you think to yourself, wow, there's a lot of different hominids.")

5. White T, et al. Pleistocene *Homo sapiens* from Middle Awash, Ethiopia. *Nature* 2003;423:Jun 12:742-747.<> (Followed by article by J. D. Clark et al on the dating of the specimen, called Herto for the region, to 157 ky. Perspective by Chris Stringer, "Out of Ethiopia," is given p692-3, including two alternative schemes for rise of moderns.)

6. McDougall I, Brown FH, Fleagle JC. Stratigraphic placement and age of modern humans from Kibish, Ethiopia. *Nature* 2005;433:Feb 17:733-736.<> (More accurate new dating of fossils called Omo I and Omo II found in 1967 by Richard Leakey and colleagues gives revised age 198 ± 12 ky.)

7. Jacobs Z, Roberts RG, Gailbraithe RF, et al. Ages for the Middle Stone Age of Southern Africa: Implications for human behavior and dispersal. *Science*;322:Oct31:733-737.

8. Hernandez RD, Kelley JL, et al. Classic selective sweeps were rare in recent human evolution. *Science* 2011;331:Feb18:920-924.

9. Vanhaeren M, et al. Middle Paleolithic shell beads in Israel and Algeria. Science 2006;312:Jun 23:1785-1788.<> (With perspective, p173, "First jewelry? Old shell beads suggest early use of symbols." Claim is made for beads dated to 100 ky or more. A similar recent find dated to 75 ky; see *Science*, Apr. 16, 2004, p369. Some are skep-

tical of the evidence, such as Richard Klein, who has long argued that the symbolic explosion began no more than 40 ky ago.)

10. Texier PJ, Porraz G, et al. A Howiesons Poort tradition of engraving ostrich eggshell containers dated to 60,000 years ago at Diepkloof Rock Shelter, South Africa. *PNAS* (*Proc. Nat'l. Acad. Sci.*) 2010;107:14:6180-6185.

11. Mourre V, Villa P, Henshilwood CK. Early use of pressure flaking on lithic artifacts at Blombos Cave, South Africa. *Science* 2010;330:Oct29:659-662.

12. Gibbon RJ, Granger DE, Kuman K, Partridge TC. Early Acheulean technology in the Riertputs Formation, South Africa, dated with cosmogenic nuclides. *J Hum Evol* 2009;56:2:152-160.<>

13. Balter M. South Africa cave slowly shares secrets of human culture. *Science* 2011;332:Jun10:1260-1261.

14. Wilson AC, Cann RL. The recent African genesis of humans. *Scientific American* 1992:Apr:66-73.

15. Sykes B. The Seven Daughters of Eve. New York, London: W.W. Norton & Co., 2001.<>

16. Thorne AG, Wolpof MH. The multiregional evolution of humans. *Scientific American* 1992:Apr:76-83.

17. Manica A, et al. The effect of ancient population bottlenecks on human phenotypic variation. *Nature* 2007;448:Jul 19:346-348.<> Data plotted on world maps shows clearly loss of genetic diversity with distance from Africa. Also shown is loss of phenotypic diversity, meaning diversity of outward appearances.

18. Jones D. Going global. *New Scientist* 2007:Oct 27:35-41.<> (Good summary of the current picture, with references to several of the key papers.)

19. Bower B. The human wave DUPE SEE 262. *Science News* 2005;168:Aug 6:92-92.<> (Subtitle, "People may have evolved fluidly, with lots of interbreeding." Report of work by Eswaran, Harpending and Rogers in the July issue of *J. Hum. Evol.*, described as "one of several new attempts to understand people today as genetic products of two or more ancient human-like populations that interbred at least occasionally".)

20. Stone R. Signs of early Homo sapiens in China? DUPE USE 387. *Science* 2009;326:Oct30:655.

21. Walter RC, et al. Early human occupation of the Red Sea coast of Eritrea during the last interglacial. *Nature* 2000;405:May 4:65-69.<> (For helpful perspective by Chris Stringer, see p24-7, "Coasting out of Africa.")

22. Macauley V, et al. Single, rapid coastal settlement of Asia revealed by analysis of complete mitochodrial genomes. *Science* 2005;308:May 13:1034-1036.<> (With perspective by P. Forster and S. Matsumura, p965-6, and widely discussed elsewhere, e.g. in *New Scientist*, p14, May 21.)

23. Bower B. Going coastal. *Science News* 2007;172:Oct 20:243-244.

24. Li H, Durbin R. Inference of human population history from individual whole-genome sequences. *Nature* 2011;475:Jul28:493-496.

25. Liu H, Prugnolle F, Manica A, Balloux F. A geographically explicit genetic model of worldwide human settlement history. *Am J Hum Genet* 2006;79:2:230-237.

26. Balter M. Of two minds about Toba's impact. *Science* 2010;327:Mar5:1187-1188.<> Report on meeting about Toba volcano held Feb 20, Oxford, UK. New evidence casts doubt on hypothesis that most of our species was wiped out by the climate chill caused by the giant volcanic eruption 74,000 years ago (74 kya). Touches on other issues. Sites in India indicate *H. sapiens* was there at this time, judging by tools, but was possibly Neanderthal. However this conflicts with latest mtDNA dating putting African exodus 55-70 kya (*Am. J Hum. Genetics*, June 2009). But all these issuers are controversial.

27. Marean CW. When the sea saved humanity. *Scientific American* 2010;2010:Aug:55-61.<>

28. Editor. No marathons for Neanderthals. *Nature* 2011;470:Feb27:439.<> (Brief note referring to *J. Hum. Evol.*, 60:299-308, on an anatomical study showing that the Neanderthal foot was not suited for distance running. Also noted in Science News, 3/12/11, pg 8, under caption, "Humans could outrun Neanderthals.")

References, Ch. 7.

1. Marshall M. Out of Africa, into bed with the locals. *New Scientist* 2011;2011:Jun18:11.<> (Refers to reports at a recent meeting in London of the Royal Society, on work by Peter Parham, based in part on the acquisition of novel genes for resisting local diseases, and on the Denisovan genome.)

2. Young E. Our hybrid origins. *New Scientist* 2011:July 30.<> Cites several authoritative recent reports that conflict sharply with the picture just a few years ago, and in particular, demonstrating the we do indeed harbor Neanderthal genes, and in addition, genes of at least one other archaic population.

3. Green RE, et al. A draft sequence of the Neanderthal genome. *Science* 2010;328:May7:710-725.<> (About 50 authors named, Paabo last. Informative perspective provided pg 680-4, under title, "Close encounters of the prehistoric kind.")

4. Hudjashov G, Kivisild T, Underhill PA, Endicott P, et al. Revealing the prehistoric settlement of Australia by Y chromosome and mtDNA analysis. *Proc Natl Acad Sci USA* 2007;104:21:8726-8730.<> They find "no evidence of any archaic maternal or paternal lineages in Australians" but do acknowledge unusual morphological features. Their evidence supports long isolation of the Australian gene pool.

5. Editor. Not out of Java. *Science* 2009;326:Oct30:649.<> (Short commentary on work reported in Archeology in Oceania, in which 26 Australian skulls of age 15-26 kya were analyzed compared to 19 more archaic *H. sapiens* skulls, as old as 195 kya, and to 5 *H. erectus* skulls from Java. The more recent Australian skulls were of two types, gracile and robust. The Java specimens were outside the range of either type of Australian carnia. However, at least one authority, F. Smith, is not convinced and says "the jury is still out.")

6. Durband AC. The view from down under: a test of the multiregional hypothesis of modern human origins using the basicranial evidence from Australia. *Coll Anthrop* 2007;31:3:651-659.<> Report of re-examination of cranial base features, which

tend to be highly conserved, from recent and fossil native Australians and strongly confirms their morphological uniqueness. This remains to be explained in terms of genetic analysis.

7. Curnoe D. A 150-year conundrum: cranial robustness and its bearing on the origin of aboriginal Australians. *Int J Evol Biol* 2011;2011:632484.<> Gives good historical background, Author argues that the cranial robustness of this population remains to be explained, possibly in terms of phenotypic plasticity, and urges more research in this direction.

8. Summerhayes GR, Leavesley M, et al. Human adaptation and plant use in highland New Guinea 49,000 to 44,000 years ago. *Science* 2010;330:Oct1:78-81.<> (With informative perspective by Chris Gosden, pg 41-2.)

9. Hart CWM, Pilling AR. The Tiwi of North Australia. NY: Holt, Rinehart and Winston, 1979.<> (Fieldwork edition. First edition 1960. Of the series "Case Studies in Cultural Anthropology;" George and Louise Spindler, ed's, Stanford Univ.)

10. Harris M. The Rise of Anthropological Theory. N.Y.: Thomas Y. Crowell & Co. (1st edn), 1968.<> (Updated edition appeared 2001 (Rowan & Littlefield) with introduction on this history of the book, now a classic, by M.L. Margolis. The book assumes that the reader is already familiar with the topics and authors discussed. Plenty of downside but remains a good survey and a rich source of key literature.)

References, Ch. 8.

1. Weckler JE. Neanderthal man. *Scientific American* 1957;1957:Dec:89-96.

2. Trinkhaus E, Howells WW. The Neanderthals. *Scientific American* 1979:Dec:118-133.

3. Wong K. Twilight of the Neanderthals. *Scientific American* 2009;2009:Aug:32-37.<> Note about this magazine: Until about the 1980's, this magazine invited leading authorities to present their findings and overviews of their fields. It is now written mainly by science journalists for wider popular appeal.

4. Lieberman P. Voice in the wilderness: How humans acquired the power of speech. *The Sciences* 1988:Jul/Aug:23-29.<>

5. Lieberman P. Human Language and Our Reptilian Brain. Cambridge, MA: Harvard Univ. Press, 2000.<> (Subtitle, "The subcortical bases of speech, syntax and thought." Regarding Neanderthals, discusses his hypothesis, coauthored with Crelin [*Science* ___], in light of new findings, p135-150, and see Note 5, p176-7.)

6. Finlayson C, et al. Late survival of Neanderthals at the southernmost extreme of Europe. *Nature* 2006;443:Oct 19:850-853.<> (With perspective by E. Delson and K. Harvati, p762-3, and widely discussed elsewhere such as in *Science* 313:1557 and in *Science News* 170:205.)

7. Kraus J, et al. Neanderthals in central Asia and Siberia. *Nature* 2007;449:Oct 18:898-904.

8. Slimak L, Svendsen JI, et al. Late Mousterian persistence near the Arctic Circle. *Science* 2011;332:May13:841-844.<> (With perspective, pg 778.)

9. Guthrie RD. The Nature of Paleolithic Art. Chicago: Univ. Chicago Press, 2006.<> Reviewed in *Nature* 441:575 by Paul G. Bahn. According to Bahn, the theories about why the art was done are speculative and controversial, including those in this book, but one need not agree with the theories to enjoy the high quality of this collection of many of the major works.

10. Leroi-Gourhan A. The evolution of Paleolithic art. *Scientific American* 1968;1968:Feb:58-70.<> Ranks the works of cave art into periods of increasing sophistication. Includes table of some highly stylized repeating symbols which occur widely and have been compared to herald coats of arms. Numerous books and associated theories have since been written on this subject.

11. Balter M. Going deeper into the Grotte Chauvet. *Science* 2008;321:Aug15:904-905.<> News report on further work on the marvelous cave paintings found here, discovered only recently (1994) and dated to 31,000 years. However, the paintings seem too advanced for such as old date, and there are serious questions about the accuracy of that date.

12. Ravilious K. Messages from the Stone Age. *New Scientist* 2010;2010:Feb20:30-34.<> Tells of the concept hatched by student Genevieve van Petzunger, supervised by A. Nowell, reported at the *Paleoanthropological Society* meeting (Chicago, April), to the effect that many of squiggles, hatched lines, patterns of dots, zig-zags, etc. frequently seen in ancient rock art and cave art, might actually be a kind of primitive writing.

13. Conard NJ. A female figurine from the basal Aurignacian of Hohle Fels Cave in southwestern Germany. *Nature* 2009;459:May14:248-252.<> For perspective by Paul Mellars, see pg 176-7.

14. Balter M. Seeking the key to music. *Science* 2004;306:Nov 12:1120-1122.<> Flute fashioned from bird bones, found in Germany and France, dated to 32 ky, "the oldest undisputed evidence of music."

15. Vanhaeren M, et al. Middle Paleolithic shell beads in Israel and Algeria. *Science* 2006;312:Jun 23:1785-1788.<> (With perspective, p173, "First jewelry? Old shell beads suggest early use of symbols." Claim is made for beads dated to 100 ky or more. A similar recent find dated to 75 ky; see *Science*, Apr. 16, 2004, p369. Some are skeptical of the evidence, such as Richard Klein, who has long argued that the symbolic explosion began no more than 40 ky ago.)

16. Solecki RS. Shanidar Cave. *Scientific American* 1957:Nov:58-64.

17. Hooper R. Neanderthals bid for human status. New Scientist 2007:Jun 16:12.<> (Reviews work by Terry Hopkins and colleagues appearing in *Antiquity* 81:294, arguing that Neanderthals are under-rated, showing clear signs of long-term progress, notably in merging two tool styles from Lower Paleolithic, more than 300 kybp, into a superior method, called Levallois. Also spread into areas previously too harsh and cold.)

18. Zilhao J, Angelucci DE, et al. Symbolic use of marine shells and mineral pigments by Iberian Neanderthals. *PNAS (Proc. Nat'l. Acad. Sci.)* 2010;107:3:1023-1028.

19. Balter M. Did working memory spark creative culture? *Science* 2010;328:Apr9:160-163.<> Devoted mainly to the theory of Wynn and Coolidge. Related or alternative

theories are reviewed on pages 164-7 of this special section entitled "Evolution of Behavior."

20. Wynn T, Coolidge FL. A stone-age meeting of minds. *American Scientist* 2008;96:JanFeb:44-51.

21. Rimol LM, Agartz I, et al. Sex-dependent association of common variants of microcephaly genes with brain structure. *PNAS (Proc. Nat'l. Acad. Sci.)* 2010;107:1:384-388.

22. Vallender EJ, Mekel-Bobrov N, Lahn BT. Genetic basis of human brain evolution. *Trends Neuroscience* 2008;31:12:637-644.

23. Gilbert SL, Dobryns WB, Lahn BT. Genetic links between brain development and brain evolution. *Nature Reviews Genetics* 2005;6:7:581-590.

24. Beja-Pereira A, Luikart G, et al. Gene-culture coevolution between cattle milk protein genes and human lactase genes. *Nature Genetics* 2003;35:4:311-313.<>

25. Laluez-Fox C, al e. A melanocortin 1 receptor allele suggests varying pigmentation among Neanderthals. *Science* 2007;318:Nov 30:1453-1455.<> (Commentary by E. Culotta appeared in the News section of issue of Oct. 26. By coincidence, this issue also gives new insight into dog coat color, p1418, with perspective p1395.)

26. Gibbons A. European skin turned pale only recently, gene suggests. *Science* 2007;316:Apr 20:364-365.<> (Reporting highlights of meeting of *Amer. Assoc. Phys. Anthropol.*, Philadelphia, Mar. 28-31.

27. Culotta E. Ancient DNA reveals Neanderthals with red hair, fair complexions. *Science* 2007;318:Oct 26:546-547.<> (Perspective on report online, to be published in upcoming issue.)

28. Staff. Selective power of UV. *Science* 2000;289:Sep1:1461.<> Note in the "Random Samples" column of a study by anthropologists Nina Jablonski and George Chaplin of satellite data on UV radiation world-wide, as correlated with skin color of people living at various latitudes. They conclude that this "bolster's the case" for solar radiation as the main cause of skin color. Findings published in *J. Hum. Evol.*, July.)

29. Pennisi E. The dawn of stone age genomics. *Science* 2006;314:Nov 17:1068-1071.

30. Hodgson JA, Disotell RE. No evidence of a Neanderthal contribution to modern human diversity. *Genome Biology* 2008;9:206:1-7.

31. Bower B. Ancient gene yield. *Science News* 2006;170:Nov 18:323.<> (Says that an upcoming article in PNAS by Bruch T. Lahn supports interbreeding, possibly with Neanderthals, since a recently acquired gene dated to 37 ky is now in 70% of humanity, and likely came from another line. Geneticist M.F. Hammer remarks that it "adds to growing genetic evidence of inbreeding among various lines of human ancestors, within and outside of Africa." Editorial by D.M. Lambert and C.D. Millar says full Neanderthal genome expected in near future. Rubin dates Neanderthal split back to 370 ky.)

32. Clabby C. Paleogenomic puzzles. *American Scientist* 2011;99:May/Jun:210.<> (Commentary on the several puzzles arising from the recently genetic studies showing archaic inputs to modern humans.)

33. Staff. Patrimony debate gets ugly. *Science* 1999;285:Jul9:195.<> (Item in *Random Samples* column: "The Neanderthal wars erupted again late last month over a child buried in Portugal 25,000 years ago ... whether or not Neanderthals were a

separate species is still a hot-button issue." The fossil shows mixture of features of Neanderthals and moderns. Lots of nasty name-calling among top authorities. This is but one example of many. By the way, on the same page is reported an "impossible" living rat since it has double the normal number of chromosomes, showing again how limited is our knowledge of genetics.)

34. Bower B. Asian trek. *Science News* 2007;171:Apr 7:211.<> (Tells of H. Shang et al, writing in upcoming issue of *Proc. Nat'l Acad. Sci.* [PNAS] finding remains in China 40,000 years old with features combining those of Neanderthals and so-called archaic traits, suggesting interbreeding. A possibly related skeleton was found in China in 1958, discussed by G. Barker and colleagues in the March issue of *J. Hum. Evol.*)

35. Staff. Skull and crossed bones. *Nature* 2006;444:Nov 9:126.<> (Note on work by Eric Trinkaus and colleagues in *Proc. Nat'l. Acad. Sci. USA*, of new analysis of Neanderthal bones dated 30 ky from a Romanian cave, showing they had a mixture of features of classic Neanderthal and later modern humans. Authors claim two other sets of Neanderthals also show mixed features and conclude interbreeding went on.)

36. Wong K. Out with a bang. *Scientific American* 2010;2010:Dec:26.<> (Summary of work from St. Petersburg, Russia, by L.V. Goluvanova and colleagues, appearing in *Current Anthropology*, reporting a thick layer of ash found in a cave site dated to 40 kya, due to a major eruption of a volcano in the Caucasus at that time.)

37. Neves AGM, Serva M. Extremely rare interbreeding events can explain Neanderthal DNA in modern humans. *Cornell Univ. Library, online: arxiv.org/abs/1103.4621.* 2011:Jun6:1-20.<> (Assumptions appear to be based on 130,000 year history of alternating or mixed African and Neanderthal populations in the caves of Skuhl and Kafzeh, in Israel; and the observed 1% to 4% Neanderthal DNA in modern humans. They conclude that this could account for the latter's extinction, provided the Neanderthals were not inferior.)

38. Buchanan M. Neanderthals may have drifted gently into oblivion. *New Scientist* 2011;2011:Apr9:11.<> (News report of study by Neves and Serva, with comments from other experts queried, who give a favorable opinion but raise some caveats.)

39. Mellars P, French JC. Tenfold population increase in Western Europe at the Neanderthal-to-modern human transition. *Science* 2011;333:Jul29:623-627.<> (With informative perspective by Bouquet-Appel, pg 560-1.)

40. Banks WE, d'Errico F, etal. Neanderthal extinction by competitive exclusion. *PloS ONE* 2008;3:12:e 3972-.

41. Gibbons A. Modern humans made their point. *Science* 2005;308:Apr22:490-491.<> Reports highlights of meeting of paleoanthropologists, this item being lecture by Shea and Brooks arguing that modern humans likely possessed the bow-and-arrow, giving them a huge advantage over Neanderthals. Other items reported: possible archaic genes in modern people; early inhabitants of England; genetic differences among ethnic populations world-wide; and studies on some living early primates, the lorises of Madagascar.

42. Staff. Did early hunters get a head start? *New Scientist* 2009;2009:Jun20:14.<> Note on work by Matthew Sisk and John Shea who argue that very early stone blades could

have, and possibly or probably did, make effective arrowhead as early as 100,000 years ago. Published in *J. Archeol. Sci.*, DOI: 10.1016/j.jas.2009.05.023.

43. Rosas A, Martinez-Maza C, et al. Paleobiology and comparative morphology of a late Neanderthal sample from El Sidron, Asturias, Spain. *PNAS (Proc. Nat'l. Acad. Sci.)* 2006;103:51:19266-19271.

44. Fabre V, Condemi S, Degioanni A. Genetic evidence of geographical groups among Neanderthals. *PLoS ONE* 2009;4:4:e 5151.

45. Gravina B, Mellars P, Ramsey CB. Radiocarbon dating of interstratified Neanderthal and early modern human occupations at the Chatelperronian type-site. *Nature* 2005;438:Nov3:51-56.<> Occupation of this site appears to have alternated between moderns and Neanderthals, from about 41 to 43 ky ago. Site was discovered in 1840's. Evidence is alternating strata of Chatelperronian and Aurignacian (early modern human) stone tool styles, but Mousterian (classical Neanderthal) were absent.

46. Zilhao J, d'Errico F, Bordes JG, Lenoble A, Texier JP, Rigaud JP. Analysis of Aurignacian interstratification at the Chatelperronian-type site and implications for the behavioral modernity of Neanderthals. *PNAS (Proc. Nat'l. Acad. Sci.)* 2006;103:33:12643-12648.<> Authors interpret this site to imply that Neanderthals were copying the tool styles of the modern humans.

47. Mellars P, Gravina B, Ramsey CB. Confirmation of Neanderthal / modern human interstratification at the Chatelperronian type-site. *PNAS (Proc. Nat'l. Acad. Sci.)* 2007;104:9:3657-3662.

48. Bower B. French site spurs Neanderthal debate. *Science News* 2005;168:Sep 17:189.<> (Short summary of work led by P. Mellars in upcoming *Nature* that before dying out about 28 kybp, the Neanderthals borrowed tool styles of neighboring modern humans. This is based on their study of Neanderthal remains in a cave in France showing two periods of occupancy, 40-39 kybp and 36-34.5 kybp, with tool style dubbed Chatelperonnian, which they argue shows features seen in tools of the later modern humans. However, another expert, Joao Zilhao, believes the modified tools were independently developed by the Neanderthals.)

49. deLeon MSP, Golovanova L, Doronichev V, Romanova G, Akazawa T, Condo O, Ishida H, Zollikofer CP. Neanderthal brain size at birth provides insights in the evolution of human life history. *PNAS (Proc. Nat'l. Acad. Sci.)* 2008;105:37:13764-13768.<> From the concluding paragraphs: "It could be argued that growing smaller, but similarly efficient, required less energy investment and might ultimately have led to higher net reproduction rates".

50. Gibbons A. Dental evidence suggests Neanderthals matured faster than we do. *Science* 2007;318:Dec7:1547.<> News item on work by B. Holly Smith. Articles in several sources are discussed. See also pg 1646, "Paleontologists get X-ray vision," which touches on the same matter.

www.ingramcontent.com/pod-product-compliance
Lightning Source LLC
Chambersburg PA
CBHW051546170526
45165CB00002B/902